EUCALYPTS OF SOUTH AUSTRALIA

Eucalyptus kingsmillii subsp. *alatissima*

Dean Nicolle

First Published in 1997 by
Dean Nicolle
156 Pimpala Road, Morphett Vale S.A. 5162

National Library of Australia
Cataloguing-in-Publication data

Nicolle, Dean, 1974 -
Eucalypts of South Australia
includes index
IBSN 0 646 32743 7

1. Eucalyptus - South Australia - Identification. I. Title.

583.766099423

Printed in South Australia by Lane Print Group

CONTENTS

E. pimpiniana

E. dalrympleana subsp. *dalrympleana* near Bridgewater, Mount Lofty Ranges, S.A.

ACKNOWLEDGMENTS

Firstly I would like to acknowledge a grant from the Queen's Trust which funded (in part) the production of this book.

I am grateful to the Anangu Pitjantjatjara people for permission to travel on their lands to see *E. gillenii* and especially to David Umala for accompanying me on two field trips on their lands. I am also grateful to the Maralinga Tjarutja people, particularly Archie Barton A.M. for permission, on many occasions, to travel on their lands.

Many people have helped in discussions on eucalypts in general and in accompanying me on field trips, particularly John Conners, Peter Lang and Malcolm French. John Jessop and staff at the State Herbarium, especially Martin O'Leary, are thanked for the use of herbarium facilities.

Other people have helped with certain species, I thank Neville Bonney for bringing to my attention and help with the species at Carpenter Rocks; Ian Wright for permission to use the soon to be published name *E. diversifolia* subsp. *hesperia*; Peter Davis for access and transport to and from Boston Island and Ivan Holliday for bringing to my attention the type locality of *E. orbifolia.*

Special thanks to Ian Brooker for his ongoing discussions on all aspects of eucalypt taxonomy and for checking much of the manuscript of this book. Finally, I thank my family for their ongoing patience and support of this project.

E. cosmophylla, an unusual form with white flowers fading to pink.

Introduction

Eucalypt is the common name applied to any plant of the genus *Eucalyptus* (the gums, mallees, stringybarks, boxes etc.), and more recently to *Corymbia* (the bloodwoods) and *Angophora* (the apples). Of the eucalypts described here, 93 are from *Eucalyptus* with the remaining two species belonging to *Corymbia*. No living *Angophora* species are known to occur naturally in South Australia.

The eucalypts dominate most South Australian landscapes, being absent only from some parts of the Nullarbor Plain and from some of the most arid areas in the north of the State. The vast diversity of eucalypt species in South Australia is perhaps a reflection of their occurrence on almost every landscape in the State. From the dry, sandy desert regions in the north to exposed coastal cliffs on Kangaroo Island to the swampy flats of the State's South-east, virtually every habitat is home to at least one species of eucalypt. They range in form from low, shrubby mallees to tall, erect forest trees and their range of bark types, leaf forms and bud, flower and capsule colour and shape is just as diverse. Flower colour, for example, ranges from white to bright yellow, pink, red and orange to even purple.

The great diversity of eucalypt species that have adapted to almost every environment in South Australia is the reason why so many species are useful and interesting. Many are valuable in the production of constructional and feature timbers and fuelwood, depending on the species; pulp; the production of honey, medicinal and perfumery oils and tannins; the use of both juvenile and adult leaves, buds, flowers and fruits for the floral and craft industry; the control of wind and water-caused soil erosion, screening and windbreaks. Eucalypts are also valuable because of the high ornamental appeal of many species and their ability to enhance the environment through shade, shelter, pollution control and the creation of animal habitats.

This book includes all the eucalypt species, subspecies and varieties known to occur naturally in South Australia. A total of 95 eucalypts are described in full (79 species and 16 subspecies and varieties), 24 of which are endemic to the State (occur nowhere else in the world). Ten of the eucalypts described in this book are not yet formally named. In the appendix a further 11 species are briefly described which although not known to occur naturally in South Australia, occur close to the border and may be undiscovered in the State.

Some of the species included are rare and the conservation of these species in the wild and through cultivation is critical. *E. paludicola, E. wyolensis* and *E.* species

'Carpenter Rocks' are examples of very rare species endemic to South Australia. Other species are common interstate but have a very minor occurrence in South Australia, such as *E. bicostata, E. gillenii, E. orbifolia, E. pauciflora* and *E. leptopoda*.

Many of the eucalypts described in this book are recently discovered or recognised. There is no doubt that more South Australian species will be recognised and perhaps discovered in the future.

Eucalyptus X '*desquamata*', see page 164

How to use this book

This book is intended to assist in the field identification of any native South Australian eucalypt. The book is designed for use by anyone with an interest in knowing what plant they are looking at, and how to grow or use it, and not just for those with prior botanical knowledge.

The species are arranged in taxonomic order, an order which places each species nearest to its closest relatives. This means that closely related and therefore usually similar species are placed on adjacent pages, allowing easy comparison. An index of both scientific and common names, in alphabetical order, is included.

Under the scientific and common name(s) for each species in the digest, the taxonomic classification for each species is given. Related species are placed in the same series (say the S.A. blue gums, series *Melliodorae*); related series are placed in the same section (say all the boxes, section *Adnataria*) and related sections are placed in the same genus (or genus equivilent, in this case *Symphyomyrtus*). This enables groups of closely related species to be identified. Some groups of related eucalypts share common features that are useful in identifying them from other groups but cannot be used to distinguish different species within that group. For example, all species of series *Rufispermae* have glossy, reddish seeds and no other eucalypt series possess such seeds. Similarly, species from section *Bisectaria* series *Accedentes* and section *Dumaria* as well as occasionally in section *Adnataria* possess pith glands (small dark spots in the pith of the branchlets) whereas all other species in this book never have pith glands.

The digest for each of the species, subspecies and varieties is divided into five informative sections:

NAME : explains the meaning of the scientific name, often helpful when learning and recalling the names of eucalypts.

DESCRIPTION : describes the conspicuous and important features of the species, including plant form (habit); bark, leaf, bud and fruit type and flower colour. Flowering time is only given when is is consistent for a given species and reliably known. Many desert species flower irregularly, such as after high rainfall events. Other feature such as the branchlets, juvenile leaves and seeds are described if considered unusual or if required for identification. Non-botanical terms are used wherever possible, however, a glossary is included for unfamiliar botanical terms where no alternative exists.

DISTRIBUTION & HABITAT : describes the natural distribution of the species and the general environment, vegetation and soil type in which the species naturally grows. Commonly associated eucalypt species are listed.

CULTIVATION & USES : mentions how common the species is in cultivation, what environmental and climatic conditions the species prefers and other information relevant to its cultivation. Uses for the species are also given and if the species is not known in cultivation suggestions are given.

NOTES : outlines any interesting facts about the species. Closely related species and how to distinguish them is also given.

A map of South Australia for each species, subspecies and variety is given, showing its distribution within the State. These have been constructed as a result of many extensive field trips, in conjunction with selected herbarium data. There is always some chance a species nay be found at a new locality outside the distribution shown on the map.

All habit and bark illustrations are from naturally occurring plants, usually photographed in South Australia. The location of the habit illustration(s) is included for interest. Illustrations of leaves, buds, flowers and fruits have come both from native plants and from plants growing in cultivation, mostly from Currency Creek Arboretum.

E. ceratocorys

Corymbia eremaea (D.J. Carr & S.G.M. Carr) K.D. Hill & L.A.S. Johnson subsp. *eremaea*

Range Bloodwood; Hills Bloodwood; Narrow-leaved Bloodwood

Genus: *Corymbia*; section: *Rufaria*; series: *Dichromophloiae*

NAME : From the Greek *eremia* meaning 'desert', referring to the occurence of this species in arid areas.

DESCRIPTION : A small tree 5 to 8 metres tall, sometimes multi-stemmed and mallee-like. The bark is rough, tessellated or flaky and red-brown to grey. Adult leaves are dull to slightly glossy, light green and lacking an intramarginal vein. The buds and fruits occur in large, terminal clusters on the outside of the crown. The flower buds are smooth (scurfy in *C. terminalis*) and open to reveal white flowers. The woody, urn-shaped fruit (9 to 22 mm long by 9 to 22 mm wide) on slender pedicels (stalks) are shed annually after seed release.

DISTRIBUTION & HABITAT : The common bloodwood of north-western South Australia, extending into adjacent hilly areas of Western Australia and Northern Territory. It favours the foothills and around the base of rocky rises and mountains which are locally moister than surrounding areas due to the run off from rain. It is less common on sandy plains and low sand dunes near hills and ranges. It is common around the Everard, Birksgate, Mann and Musgrave Ranges. It occurs southwards into the Great Victoria Desert around the Birksgate Range, always within sight of rocky hills. Commonly associated species include *E. intertexta* and *E. sparsa*.

CULTIVATION & USES : Rare in cultivation, trials have shown it to be slow growing in the south of the State where it eventually forms a small bushy tree. An attractive bloodwood worthy of trying in hot areas in the north of the State. Seedlings are susceptible to root rot in wet soils.

NOTES : Easily distinguished from *C. terminalis*, the only other bloodwood known in South Australia, by its finer overall appearance; smaller leaves, buds and fruits on slender peduncles and pedicels; the smooth buds and the north-western distribution. *C. eremaea* was previously included in *E. dichromophloia*, (now *C. dichromophloia*) a smooth-barked bloodwood endemic to tropical parts of Northern Territory.

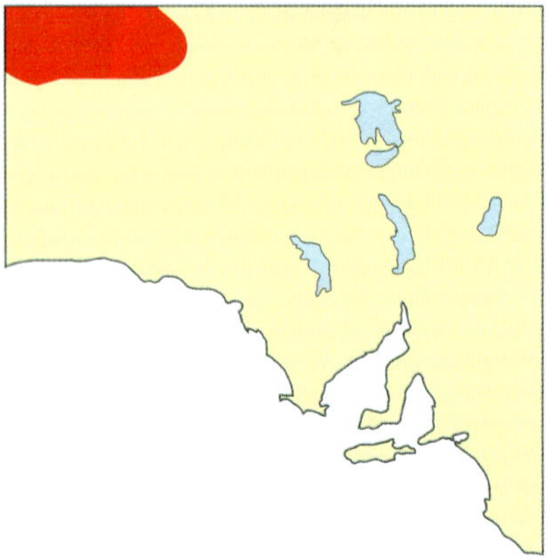

C. eremaea subsp. *eremaea* near Ronald Well in the Everard Ranges, S.A.

Corymbia terminalis (F. Muell.) K.D. Hill & L.A.S. Johnson

Plains Bloodwood; Bloodwood

Genus: *Corymbia*; section: *Rufaria*; series: *Dichromophloiae*

NAME : From the Latin *terminalis* meaning 'terminal', referring to terminal clusters of buds, flowers and fruits on the outside of the crown.

DESCRIPTION : A small tree to 12 metres tall, often of poor form and sometimes multi-stemmed. The bark is rough to the small branches, tessellated, flaky, red-brown to grey. Adult leaves are slightly glossy and light green. The buds are scurfy (finely rough) and occur in umbels of seven which together form larger terminal clusters on the outside of the crown. The fruits are urn-shaped and longer than wide (16 to 35 mm long by 12 to 23 mm wide). The flowers are white, occurring usually in winter.

DISTRIBUTION & HABITAT : Restricted to the far north-east of the State in an area north of Innamincka to Cordillo Downs Station and in the Toolachee area where it occurs in scattered pure stands on plains of red sand. It is more widespread in adjacent areas of Queensland and New South Wales.

CULTIVATION & USES : Unknown in cultivation in South Australia. It would presumably require very well drained soils in the warmer parts of the State. The flowers occur in masses and are quite conspicuous. In suitable conditions it would make a fine shade tree.

NOTES : *C. terminalis* is easily distinguished from *C. eremaea* subsp. *eremaea* (the other S.A. bloodwood) by its larger, broader leaves; the larger scurfy buds and the larger fruits that are much longer than wide. All South Australian populations of this bloodwood have recently been described as the new species *C. tumescens*, however, differences between the two species are very slight and the two intergrade over large areas of Queensland and as such *C. tumescens* is here regarded as a synonym of *C. terminalis*.

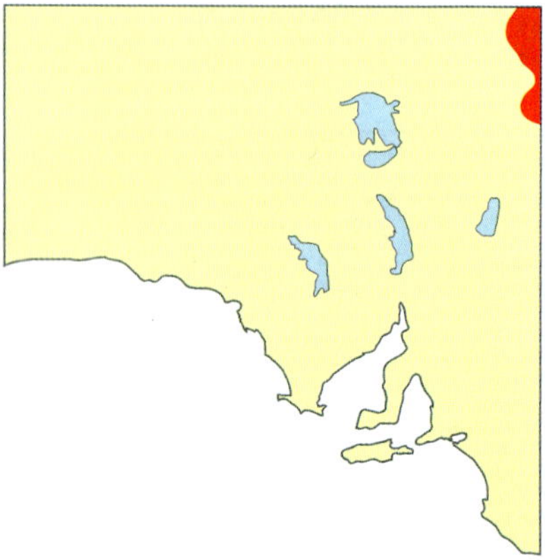

C. terminalis on the Innamincka to Cordillo Downs road, S.A.

Eucalyptus gamophylla F. Muell.

Blue Mallee; Twin-leaf Mallee; Warilu

Genus equivalent: *Eudesmia*; section: unnamed 1; series: *Odontocarpae*

NAME : From the Greek *gamos* meaning 'united' and *phyllon* meaning 'leaf', referring to the connate leaves of this species.

DESCRIPTION : A mallee, often scraggy, 1 to 4 metres tall and of equal or greater spread. *E. gamophylla* appears more dense and attractive in the few years following wildfire due to vigorous growth produced from ground level. The bark is completely smooth, or rough on the lower stems. This mallee has a crown of dull, blue-grey and glaucous leaves, with leaf pairs being connate (joined to each other around the stem). The buds and fruits are usually glaucous and occur in umbels of three, occurring usually in large clusters towards the ends of the branchlets. Small sepals (teeth-like structures) may be seen on the buds and these are retained on the fruit, which are much longer than wide. The flowers are white.

DISTRIBUTION & HABITAT : Blue mallee often forms extensive tracts of mallees on plains in deep red sands in a broad strip from the Hamersley Range in Western Australia to south of Mt Isa in far western Queensland, just extending into South Australia in the far north-west, mainly around the Tomkinson and Mann Ranges and a little way south to Mt Moulden. It often grows in pure stands, but may be associated with *E. oxymitra, E. eucentrica* and *E. mannensis.*

CULTIVATION & USES : Seldom tried in cultivation, however, trials show it to be slow growing although some seedlings have flowered prolifically only one year after planting and at only 40 cm in height. It requires a very well drained soil and a hot, dry position. Because of its ornamental bluish foliage, it is worth trying in hotter areas of the State or as a large container plant.

NOTES : The blue-grey leaves of this species make it conspicuous, even from a distance. *E. gongylocarpa* and *E. oxymitra* also have bluish leaves and may occur with *E. gamophylla* but do not have connate leaf pairs. The buds and fruits of *E. gamophylla* are also distinctive.

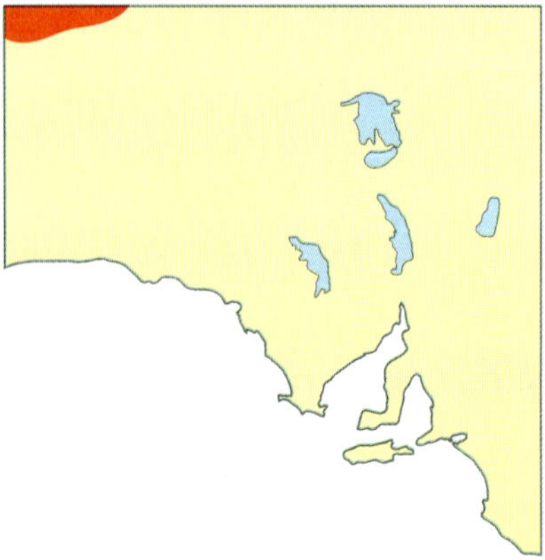

E. gamophylla on the Newman to Port Hedland road, Hamersley Ranges, W.A.

Eucalyptus gongylocarpa Blakely

Desert Gum; Marble Gum; Baarla

Genus equivalent: *Eudesmia*; section: unnamed 2 ; series: unnamed

NAME : From the Greek *gongylos* meaning 'round' and *carpos* meaning 'fruit', referring to its distinctive, spherical fruits.

DESCRIPTION : Desert gum is remarkable for the great size it attains in its arid environment, occurring as a tree 6 to 15 metres tall with a broad, shady crown. All other vegetation associated with desert gum is much lower and shrub-like. The tall habit (in relation to surrounding vegetation) and smooth white bark with small adherent red-brown flakes is characteristic. The adult leaves are dull and grey-green and juvenile leaves which are round and grey are often present in the lower crown. The small round fruits give this species one of its other common names of marble gum. The stems, leaves and fruit are usually glaucous. The flowers are white and inconspicuous.

DISTRIBUTION & HABITAT : Desert Gum occurs on the red sand dunes and intervening swales in the remote central part of the Great Victoria Desert in Western and South Australia. Here it is often the dominant tree although it is commonly associated with mallee eucalypts such as *E. glomerosa* and *E. youngiana*. It is also scattered in the south-western part of the Northern Territory and as far west as Sandstone in Western Australia.

CULTIVATION & USES : Occasionally tried in the Adelaide area with poor success and very slow growth rates, however, it is often cultivated in central Australia (Alice Springs and Yulara) with excellent results. Worthy of trying in the more arid parts of South Australia because of its attractiveness and tall form. It requires a very well drained soil in a hot, sunny position.

NOTES : A beautiful and distinctive species not likely to be confused with any other species in South Australia. Unfortunately few people have had the opportunity to see this species in its natural habitat in South Australia.

E. gongylocarpa south of Mount Poondinna, Great Victoria Desert, S.A.

Eucalyptus cladocalyx F. Muell.

Sugar Gum

Genus equivalent: *Symphyomyrtus*; section; unnamed; series: *Microcorythae*

NAME : From the Greek *clados* meaning 'branch' and *calyx* (lower part of the bud) probably referring to the flowers that occur on older twigs inside the crown.

DESCRIPTION : A large tree 12 to 30 metres tall in the Flinders Ranges and on Kangaroo Island, but a smaller more spreading tree to 15 metres tall on Eyre Peninsula (This is the cultivar known as *E. cladocalyx* 'Nana'). The tree has a colourful, smooth bark that sheds in plates giving mottled orange, yellow, grey and cream colours. The crown is terminal when mature, giving it an umbrella or storied look. Juvenile leaves are rounded and green. Sugar gum is the only South Australian native eucalypt with discolorous adult leaves (glossy, dark green upper surface, dull, paler lower surface) The elongated buds and fruits occurring inside the crown are also unique. The buds become cream before opening to white flowers in summer.

DISTRIBUTION & HABITAT : Only native in South Australia, known from four disjunct areas. In the Flinders Ranges from Wirrabara to Dutchmans Stern, occurring on the ridges, slopes and around the bases of mountains, often in rocky soils. On Kangaroo Island it often grows in pure stands forming large forests. It also occurs in hilly areas around Cleve and on lower Eyre Peninsula (Koppio Hills; Marble Range) where it commonly grows with *E. petiolaris* and *E. odorata*. Endemic to South Australia.

CULTIVATION & USES : Widely planted in Australia, especially common in southern Victoria and South Australia. It will grow in most soils with a rainfall greater than 400 mm annually. Sugar gum grows rapidly and is suitable for shade and woodlots, but is not recommended for windbreaks and wind erosion control due to its high, terminal crown.

NOTES : An easily recognisable tree with no known close relatives. Despite its restricted natural distribution, it is successfully grown over much of southern Australia and has become naturalised in some areas (e.g. Mt Lofty Ranges.) The juvenile leaves are reportedly poisonous to stock.

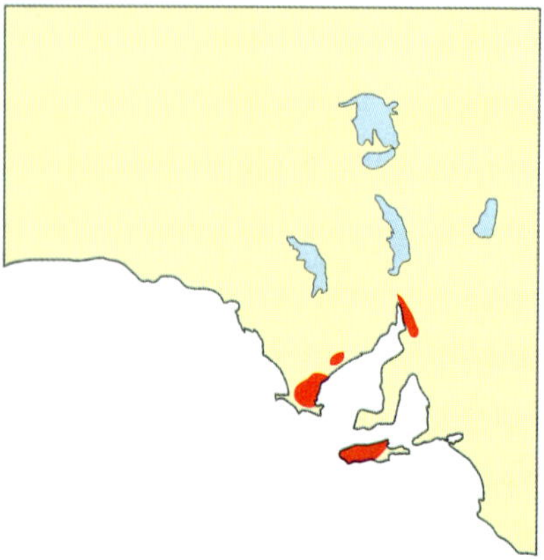

E. cladocalyx at Devils Peak, southern Flinders Ranges, S.A.

Eucalyptus paludicola D. Nicolle

Marsh Gum

Genus equivalent: *Symphyomyrtus*; section: unnamed; series: *Incognitae*

NAME : From the Latin *paludosus* meaning 'marshy' and *cola* meaning 'dwelling', referring to its usual habitat.

DESCRIPTION : Marsh gum usually develops into a tree 5 to 12 metres tall, often with a crooked trunk. It often grows in small dense thickets that together form a dense canopy. The bark is variable, usually rough and fibrous on mature trees, but it can be somewhat smooth and grey to cream, especially on younger trees. The adult leaves are green to slightly blue-green. The buds and fruits occur in umbels of seven on individual pedicels (non-sessile). The flowers are white and flowering occurs, often prolifically, in spring.

DISTRIBUTION & HABITAT : Marsh gum, as the name implies, occurs in low, seasonally swampy areas south of Adelaide and on Kangaroo Island. It is an endangered species, known from only about 25 populations, most of which occur in the Tooperang to Nangkita area in sandy soils in broad gullies. Other populations occur in the Waitpinga area and in Kelly Hill Conservation Park on Kangaroo Island. Associated species include *E. ovata, E. leucoxylon* subsp. *leucoxylon* and *E. cosmophylla.* Endemic to South Australia.

CULTIVATION & USES : A newly discovered species only recently tried in cultivation, it looks promising for use in swampy sites where few other eucalypts will grow, although trials show it will also tolerate somewhat drier sites. Plants have a moderate growth rate.

NOTES : Closely related to *E. cosmophylla*, differing in its taller habit; the often rough bark and its smaller non-sessile buds and fruits in groups of seven. *E. paludicola* often occurs with *E. ovata* but can be distinguished (among other things) by its much larger buds and fruit. A very rare species that is poorly conserved, and needs to be managed in the wild and grown in cultivation to ensure its long term survival.

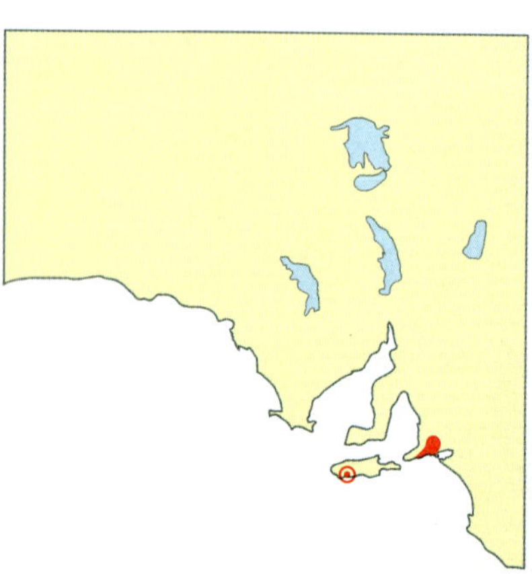

E. paludicola near the Mount Compass to Goolwa road, Fleurieu Peninsula, S.A.

Eucalyptus cosmophylla F. Muell.

Cup Gum; Bog Gum; South Australian Cup Gum

Genus equivalent: *Symphyomyrtus*; section: unnamed; series: *Incognitae*

NAME : From the Greek *cosmos* meaning 'ornament' and *phyllon* meaning 'leaf', referring to the large ornamental leaves.

DESCRIPTION : On poor stony sites, cup gum is a small unattractive mallee up to 2 metres tall, however, on more favourable sites, especially on Kangaroo Island, it is a robust mallee or small tree 3 to 8 metres tall with a crooked trunk and a low dense canopy. The bark is smooth and grey to cream. The large, leathery, dull green to blue-green leaves along with its large sessile (lacking individual pedicels) buds and large cup-shaped fruits in groups of three make this a distinctive species. Fruit size varies greatly, from 8 to 20 mm long by 10 to 22 mm wide. Flowers occur in profusion in autumn and winter and are white or occasionally become pink with age.

DISTRIBUTION & HABITAT : Cup gum is restricted to the Mount Lofty Ranges and Kangaroo Island. It inhabits low areas, often on sandy soils, or, especially in the higher parts of the Mount Lofty Ranges, on poor stony slopes with impeded drainage. Commonly associated species include *E. leucoxylon, E. baxteri* and *E. fasciculosa.* Endemic to South Australia.

CULTIVATION & USES : Occasionally cultivated as an ornamental, plants have a moderate growth rate. Cup gum will grow in sands or clays and may be useful as a windbreak in exposed sites because of its foliage to ground level or for erosion control in poor soils. Cup gum can be very ornamental, especially in flower, and would be useful as a street tree or for large gardens and parks.

NOTES : A distinctive species because of its large, thick leaves and large buds and fruits in groups of threes. It is most closely related to *E. paludicola*, distinguished by its poorer form; the thick, broader adult leaves and the larger, sessile buds and fruits in groups of threes instead of sevens.

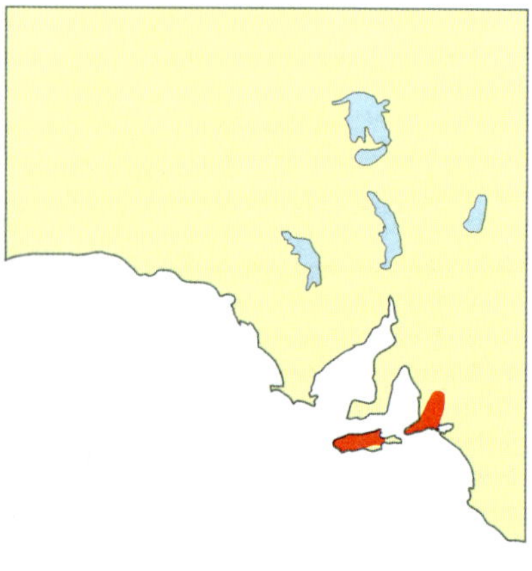

E. cosmophylla in Cox Scrub Conservation Park, near Ashbourne, S.A.

Eucalyptus trivalvis Blakely

Desert Mallee

Genus equivalent: *Symphyomyrtus*; section: *Bisectaria*; series: *Accedentes*

NAMES : From the Latin *tri* meaning 'three' and *valva* meaning 'valve', referring to the usually 3-valved fruits.

DESCRIPTION : A mallee, often of spreading habit, or a small straggly tree to 6 metres tall. The bark is rough for about 2 metres, thick, loose and grey, then smooth and light coppery-pink to brilliant glossy pink. Glands are present in the pith of the twigs. The adult leaves are dull and light green to bluish. Buds and fruits are in groups of seven. Fruits have 3 or rarely 4 valves. The flowers are white.

DISTRIBUTION & HABITAT : Not common in South Australia, occurring in scattered populations in the far north west of the State, such as near the Tomkinson and Everard Ranges. It is very scattered in the eastern Great Victoria Desert known from near Mt Igy, at Malbooma Rocks, on Mount Finke, near Lake Everard, south of Lake Gairdner and in the western part of the Gawler Ranges. It grows on or immediately below rocky hills and occasionally on gravelly rises between dunes. It also grows close to the South Australian border in Western Australia near the Serpentine Lakes and is likely to occur in South Australia in adjacent areas. *E. trivalvis* also extends to Queen Victoria Spring and the Pilbara region in Western Australia and into the Macdonnell Ranges of Northern Territory.

CULTIVATION & USES : Rarely cultivated, it is relatively slow growing in cultivation. It requires a well-drained soil and prefers hot, dry areas. Should be trialled more in the northern half of the State for use on rocky sites.

NOTES : The distribution of this species in South Australia may be more continuous and widespread than the map shows, however it appears to be rare and of unusually disjunct distribution. It is not closely related to any other eucalypt in South Australia. *E. trivalvis* is often noticeable because of its very colourful smooth upper bark and its usually bluish crown of leaves. Otherwise it is a typical desert mallee and easily ignored, with small buds and fruits.

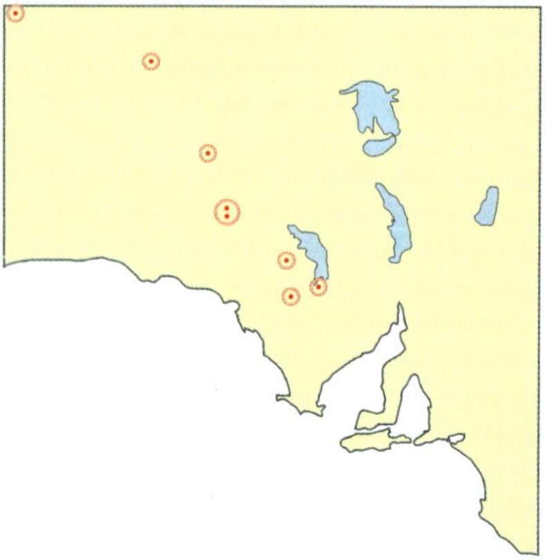

E. trivalvis at the base of Mount Finke, Great Victoria Desert, S.A.

E. trivalvis south of McPhersons Pillar, Gibson Desert, W.A.

dried fruit

Eucalyptus mannensis Boomsma subsp. *mannensis*
Mann Range Mallee

Genus equivalent: *Symphyomyrtus*; section; *Bisectaria*; series: *Bakerianae*

NAME : Both the scientific and common names refers to the area where this species has been commonly collected, the Mann Ranges in the north-western corner of South Australia.

DESCRIPTION : A mallee 3 to 6 metres tall with erect stems. The bark is rough and flaky-stringy over most of the stems. The narrow adult leaves (9 to 18 mm wide) are very glossy and green with many parallel side veins and numerous oil glands. Buds and fruits occur in umbels of seven. The obconical fruits have a thick rim and a broad, level disc. The three or rarely four valves are slightly exserted. The flowers are white.

DISTRIBUTION & HABITAT : Occurs on deep red sand dunes and the intervening plains between Range systems in the far north-west of South Australia and adjacent parts of Northern Territory and Western Australia as far west as Sandstone. Commonly associated species include *E. eucentrica* and *E. oxymitra*.

CULTIVATION & USES : Virtually unknown in cultivation. Recent trials have shown low survival rates and very slow growth in the south of the State. Trials in hotter areas may show better results. It requires a well drained soil in a hot, dry environment. The very glossy, dark green leaves gives the mallee a fresh look that is quite ornamental. It may also be useful as a windbreak and for soil erosion control.

NOTES : The only other species with similar very glossy, green leaves in the north-west of South Australia is the unrelated *E. sparsa*, which differs (among other things) in its much broader adult leaves and barrel-shaped fruits with a thin rim. In the west of its distribution in Western Australia, *E. mannensis* subsp. *mannensis* is replaced by the very similar subsp. *vespertina*. *E. mannensis* subsp. *vespertina* grows mostly in the Shark Bay area and differs from subsp. *mannensis* in its duller, slightly narrower adult leaves.

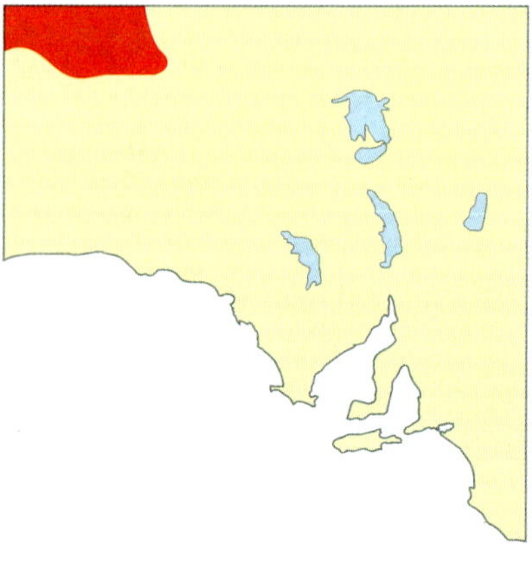

E. mannensis subsp. *mannensis* west of Neale Junction on the Anne Beadell Highway, Great Victoria Desert, W.A.

Eucalyptus cneorifolia DC.

Narrow-leaved Mallee; Kangaroo Island Narrow-leaved Mallee

Genus equivalent: *Symphyomyrtus*; section: *Bisectaria*; series: unnamed

NAME : From the genus *Cneorum* and the Latin *folium* meaning 'leaf', referring to the leaves likeness to the plant *Cneorum tricoccum*.

DESCRIPTION : On Kangaroo Island it forms dense upright mallees or slender trees to 10 metres tall. On the mainland it is more shrub-like, often with foliage to the ground. The bark is rough, hard and grey to brown on the lower half of the stems. The dense crown of erect, glossy, dark green, narrow leaves is conspicuous (leaves 6 to 10 mm wide). The small buds and fruits are in groups of seven to eleven. The fruits are sessile (individual fruits lack pedicels). White flowers appear mainly in summer and autumn.

DISTRIBUTION & HABITAT : Of restricted distribution, being common on the eastern half of Kangaroo Island, with a few small populations on the adjacent mainland near Waitpinga such as south of Parawa and on Wilson Hill. It grows in sandy loams or undulating laterite soils or less often on shallow loams over limestone. It often forms tall dense stands on Kangaroo Island. Commonly associated with *E. oleosa, E. diversifolia, E. odorata* or *E. cosmophylla.* Endemic to South Australia.

CULTIVATION & USES : Often cultivated because of its attractive foliage, however, it often grows to tree proportions when cultivated in sheltered sites and is too large for home gardens. It is moderately quick growing, and is recommended for use as a shelter, windbreak (it is bushy to the ground on exposed sites) and erosion control , especially in coastal areas and on limestone soils. The leaves of *E. cneorifolia* are used on Kangaroo Island for the production of eucalyptus oil.

NOTES : A rare species with no close relatives in South Australia (it is closely related to some Western Australian species). It is not likely to be easily confused with other species where it grows, however, *E. leptophylla* is a mallee from Kangaroo Island which also has narrow leaves, from which *E. cneorifolia* can be distinguished among other things by its rough bark and its erect, firm, darker green adult leaves.

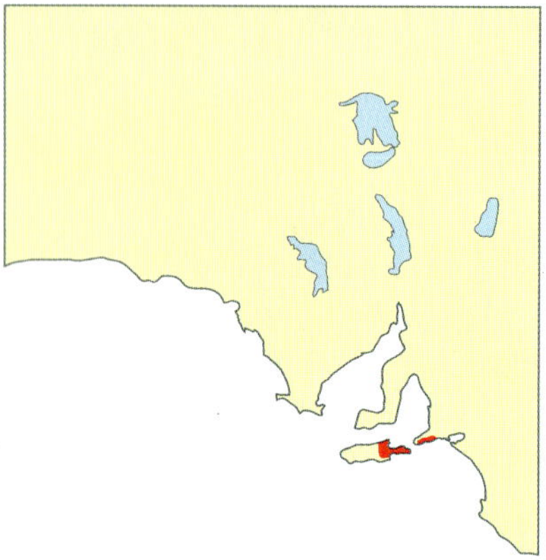

E. cneorifolia south of Parawa, Fleurieu Peninsula, S.A.

Eucalyptus oleosa F. Muell. ex Miq. subsp. *oleosa*

Red Mallee; Giant Mallee; Oil Mallee; Glossy-leaved Red Mallee

Genus equivalent: *Symphyomyrtus*; section: *Bisectaria*; series: *Subulatae*

NAME : From the Latin *oleosus* meaning 'full of oil', referring to the plentiful essential oils in the adult leaves.

DESCRIPTION : A mallee 3 to 10 metres tall. The bark is rough up to about 3 metres, hard, finely fibrous and grey, then smooth, cream and grey to coppery or sometimes completely smooth on small mallees. Pith glands are absent. The seedling leaves are very small, narrow (linear) and crowded around the stem. Adult leaves are variable in width, very glossy and green, with many oil glands when held up to light. Buds and fruits occur in umbels of seven to thirteen. The buds have a blunt, lightly ribbed operculum that is narrower than the hypanthium (egg in an egg-cup shaped buds). The fruits are rounded with fragile needle-like valves that are erect and exserted before they break off with age. The flowers are white.

DISTRIBUTION & HABITAT : Widespread and common in the settled areas of South Australia, from western Eyre Peninsula to Yorke Peninsula and throughout the Murray Mallee and upper South-east. Also on Kangaroo Island and along the coast from Waitpinga to Meningie. It occurs in a variety of soils, often over limestone, in mallee scrub with a variety of other mallee species. Extends into Victoria and New South Wales with disjunct populations in Western Australia.

CULTIVATION & USES : This mallee is not highly ornamental but is useful for growing as a windbreak and for erosion control. It is suited to most soils but will not tolerate waterlogging. Seed does not germinate as freely as some of the other mallee species. Plants have a moderate growth rate. The species is valued for honey production.

NOTES : The fine, crowded seedling leaves are distinctive in *E. oleosa*. A commonly encountered mallee distinguished by its glossy, green leaves; the blunt operculum that is narrower than the hypanthium and the small rounded fruits with fragile valves. It may be confused with *E. brachycalyx*, differing in the lack of pith glands and the more rounded fruits with inward rather than spreading valves.

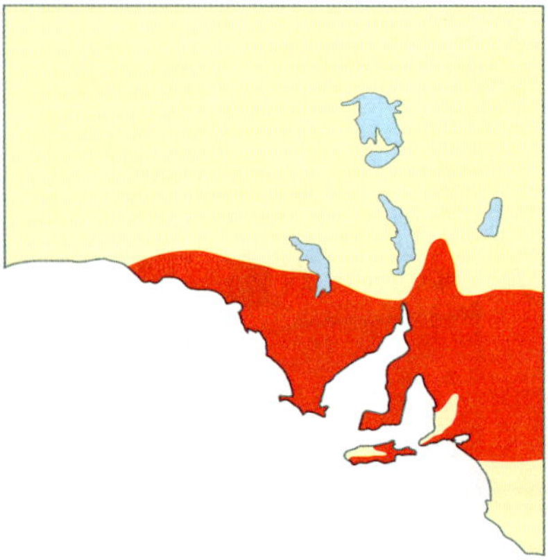

E. oleosa subsp. *oleosa* on the Princes Highway near Port Pirie, S.A.

Eucalyptus oleosa F. Muell. ex Miq. subsp. 'Desert'

Desert Red Mallee

Genus equivalent: *Symphyomyrtus*; section: *Bisectaria*; series: *Subulatae*

NAME : This is an unnamed form of *E. oleosa* restricted mainly to the Great Victoria Desert.

DESCRIPTION : A mallee 3 to 5 metres tall. The bark is rough for up to 2 metres, grey, then smooth and cream to pinkish. The juvenile and adult leaves are the same as for *E. oleosa* subsp. *oleosa*. The buds and fruits are in groups of seven to thirteen. The buds have a rounded to conical, smooth operculum that is flush with the hypanthium at the join. The fruits are rounded, with fragile, needle-like valves that are erect and exserted before breaking off with age. The flowers are white.

DISTRIBUTION & HABITAT : Of very scattered occurrence in the southern half of the Great Victoria Desert in Western and South Australia. It occurs from near Lake Gairdner westwards to at least Tjuntjunjara in Western Australia. It grows on red sand plains, sometimes with *E. concinna* and *E. eremicola*.

CULTIVATION & USES : Not known in cultivation. It would be more suited to drier, sandy soils in a hotter climate than *E. oleosa* subsp. *oleosa*. It should be very useful for windbreaks, wind erosion control and low shade in arid areas of the State.

NOTES : This unnamed subspecies of *E. oleosa* differs from the other two South Australian subspecies in its smooth, rounded to conical operculum which is equal to width to the hypanthium at the join. This subspecies and subsp. *oleosa* intergrade in an area from Port Augusta north-westwards. *E. oleosa* subsp. 'Desert' is also closely related to *E. eremicola* but differs in its fine, crowded seedling leaves; its larger, broader adult leaves and the larger buds (often with a rounded operculum) and fruits.

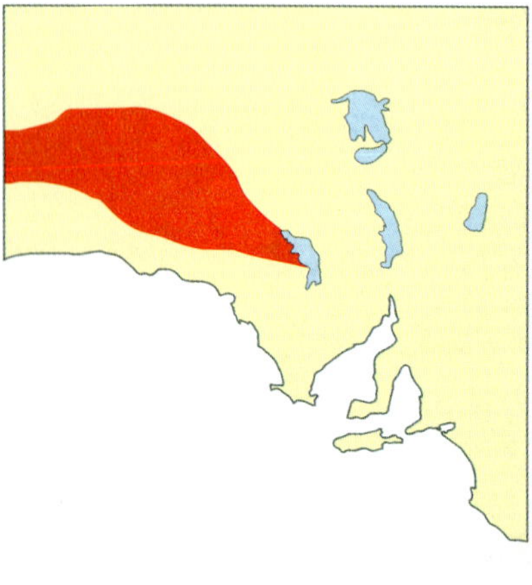

E. oleosa subsp. 'Desert' on the Tarcoola to Ceduna track (Googes Track), north of Mount Finke, S.A.

Eucalyptus oleosa F. Muell. ex Miq. subsp. 'Coastal'

Coastal Red Mallee

Genus equivalent: *Symphyomyrtus*; section: *Bisectaria*; series: *Subulatae*

NAME : This is an unnamed form of *E. oleosa* occurring in coastal regions in the west of the State.

DESCRIPTION : A mallee 3 to 7 metres tall. The bark is rough for up to 3 metres, hard but fibrous, grey, then smooth cream to grey or coppery. The juvenile and adult leaves are the same as for *E. oleosa* subsp. *oleosa*. The buds are in groups of seven to thirteen. The buds have a smooth, conical to slightly beaked operculum (similar in shape to *E. yalatensis*), that is about equal in width to the hypanthium at the join. The fruits are rounded to cupular, with fragile, needle-like valves that are erect and exserted before breaking off with age. The flowers are white.

DISTRIBUTION & HABITAT : Occurs on the edge of the Nullarbor Plain and close to the coast from about Yalata westwards into Western Australia at least as far west as Israelite Bay. It grows in loams over shallow limestone in coastal mallee scrub with species such as *E. gracilis*, *E. calcareana* and *E. yalatensis*.

CULTIVATION & USES : Uncommon in cultivation, this mallee would have the same requirements and uses as *E. oleosa* subsp. *oleosa* although is may prove more suited to exposed coastal situations.

NOTES : Differs from the other two subspecies of *E. oleosa* described here in the smooth conical to beaked operculum that is the same width as the hypanthium where they join. The exact eastern limit of this subspecies, where it would probably grade into *E. oleosa* subsp. *oleosa*, is not known. The buds and fruits of *E. oleosa* subsp. 'Coastal' are very similar to those of *E. yalatensis*, with which it may grow, but it differs in its often taller habit and glossy, green adult leaves as well as having fine, narrow, crowded juvenile leaves (broader, non-crowded juvenile leaves in *E. yalatensis*).

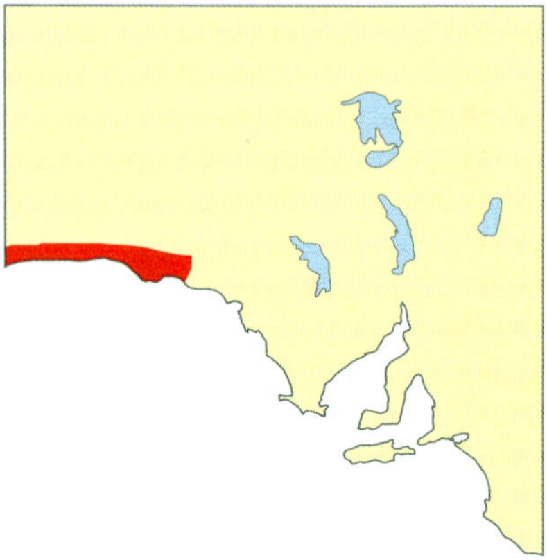

E. oleosa subsp. 'Coastal' on the Eyre Highway west of Nullarbor Motel, Nullarbor Plain, S.A.

Eucalyptus eremicola Boomsma

Vokes Hill Mallee; Nawa Mallee

Genus equivalent: *Symphyomyrtus*; section: *Bisectaria*; series: *Subulatae*

NAME : From the Greek *eremia* meaning 'desert' and *cola* meaning 'dwelling', referring to its occurrence in the Great Victoria Desert.

DESCRIPTION : A mallee, often of low, spreading habit, 2 to 5 metres tall. The bark is variable, smooth and white to grey-brown throughout or rough and brown on the lower part of the stems. Juvenile leaves are elliptical to broad-lanceolate. The adult leaves are narrow, glossy and green. The small buds and fruits occur in groups of seven to eleven. The small rounded fruits have persistent, fragile, needle-like style remnants (extensions from the valves) protruding from inside the fruits, especially when immature. The flowers are white.

DISTRIBUTION & HABITAT : An uncommon species, scattered in the Great Victoria Desert of South and Western Australia, becoming more common in the southern part of its distribution, north of the Nullarbor Plain. It occupies red sands between sand dunes, associated with species such as *E. eucentrica* and *E. youngiana*.

CULTIVATION & USES : Rarely cultivated. Suitable for the more arid areas of the State in sands or loams including alkaline soils. It is not suited to areas of rainfall more than about 450 mm annually as it lacks vigour and becomes scraggly and unattractive. It is relatively attractive with its low growth and glossy green leaves and would be suitable for low windbreaks and for parks and gardens in hotter areas of the State. It will not tolerate waterlogging.

NOTES : Closely related to *E. peeneri*, differing mainly in the glossy, green adult leaves and broader juvenile leaves. Also related to *E. oleosa* (it sometimes grows with *E. oleosa* subsp. 'Desert') differing in the broader juvenile leaves; narrower adult leaves; smaller buds with a pointed operculum and the smaller, more delicate fruits.

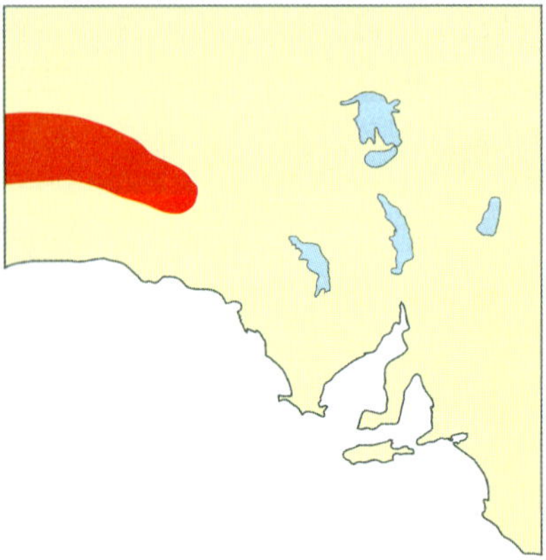

E. eremicola on the Maralinga to Oak Valley track, south-east of Lake Maurice, Great Victoria Desert, S.A.

Eucalyptus peeneri (Blakely) Pryor & Johnson ex Boomsma

Peeneri Mallee; Water Mallee

Genus equivalent: *Symphyomyrtus*; section: *Bisectaria*; series: *Subulatae*

NAME : Both the scientific and common names are from the Aboriginal name of this plant.

DESCRIPTION : Usually a low, spreading mallee under 2 metres tall, but occasionally taller. The bark is usually smooth and white to tan throughout, but some rough flaky bark is sometimes present at the base of the stems. The juvenile leaves are very narrow and bluish in colour. Adult leaves are narrow, dull, and slightly blue-green to blue-grey. The buds and fruits are very similar to *E. eremicola,* although the fruits are usually slightly larger. The flowers are white.

DISTRIBUTION & HABITAT : Scattered in the Great Victoria Desert of South Australia, extending to but becoming rare in adjacent Western Australia. It is most common in the area bounded by Vokes Hill, Emu, Ooldea and Wynbring. It grows on red sand on low sand dunes or on the intervening plains. There is sometimes limestone present in the soil.

CULTIVATION & NOTES : Rarely cultivated, trials near Adelaide show a good survival rate with *E. peeneri*, where it is slow growing but develops into an attractive many stemmed mallee about 2 metres tall by 2 metres wide. Plants flower prolifically when about 2 years old and only 50 cm tall. It may make an attractive large container plant and would be suitable for gardens because of its small size. Requires a well drained soil in a sunny position. It should be planted more for ornamental purposes in drier parts of the State.

NOTES : Closely related to *E. eremicola*, differing in the dull, bluish adult leaves; the narrower juvenile leaves and the slightly larger fruits. It is also similar to *E. yalatensis,* differing in the often smooth bark; the narrow juvenile leaves; the narrower, thinner adult leaves and the spherical fruits. Water has been obtained from the cut stems and roots of *E. peeneri* by Aborigines.

E. peeneri on the Anne Beadell Highway, east of Serpentine Lakes, Great Victoria Desert, S.A.

Eucalyptus yumbarrana Boomsma

Yumbarra Mallee

Genus equivalent: *Symphyomyrtus*; section: *Bisectaria*; series: *Subulatae*

NAME : Named from the locality where the type specimen was collected, Yumbarra Conservation Park, north of Ceduna.

DESCRIPTION : A heavily-crowned mallee, sometimes developing into a straggly tree to 8 metres tall. The bark is rough at the base or up to about a metre, loose and flaky or ribbony, then smooth, ribbony, tan to cream. The twigs, leaves, buds and fruits are always non-glaucous. The juvenile leaves are sessile, non-decurrent and green. Adult leaves are thick, glossy and dark green. The buds and fruits are in groups of seven to eleven. The flowers are cream to pale yellow.

DISTRIBUTION & HABITAT : Occurs in the far southern Great Victoria Desert of South Australia, being common in Yumbarra Conservation Park north of Ceduna. It occurs on red or orange sand on dunes and in the intervening swales in open mallee scrub, often associated with *E. ceratocorys*, *E. brachycalyx* or *E. concinna*. A form with smaller leaves, buds and fruits occurs from Maralinga to north of Cook along the Ooldea Range. Endemic to South Australia.

CULTIVATION & USES : Rarely tried in cultivation, most probably because it is rarely collected. Plants are fairly slow growing at first, but flower at a very early age and flowering is profuse when mature. It prefers sandy soils including those with limestone, in low rainfall areas and it will not tolerate waterlogging. It would be suitable for large gardens and general broad-scale plantings.

NOTES : *E. yumbarrana* is a mallee often confused with *E. socialis* and *E. eucentrica*, however, it is quite distinct with its thick, glossy, dark green adult leaves and larger non-glaucous buds and fruits. Intergrades between *E. yumbarrana* and *E. eucentrica* are rare, occurring mostly west of the Gawler Ranges. The mallee known as *E. yumbarrana* subsp. *striata* is a chance hybrid between *E. eucentrica* and *E. canescens* subsp. *canescens* collected from north of Cook.

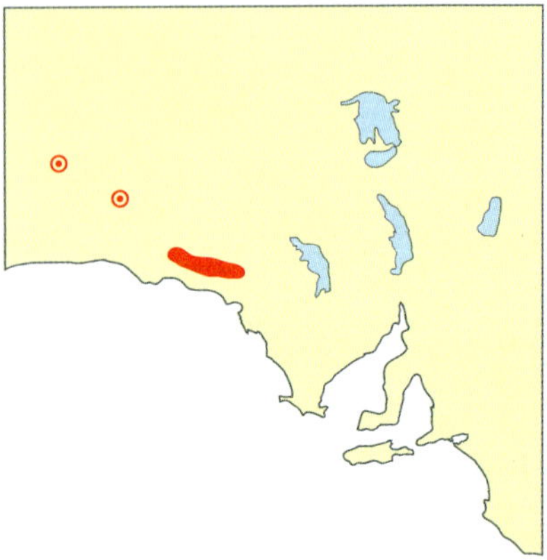

E. yumbarrana in Yumbarra Conservation Park, Great Victoria Desert, S.A.

E. yumbarrana at the junction of the Cook to Vokes Hill track and Oak Valley to Tjuntjunjara track, Great Victoria Desert, S.A.

Eucalyptus eucentrica L. Johnson & K. Hill

Inland Red Mallee

Genus equivalent: *Symphyomyrtus*; section: *Bisectaria*; series: *Subulatae*

NAME : From the Greek *eu* meaning 'well' and *kentrikos* meaning 'of the centre' referring to its occurence in the interior of Australia.

DESCRIPTION : A mallee 4 to 7 metres tall. The bark is rough for 2 to 3 metres, thick and loose, grey to tan, then smooth, cream. The twigs are often glaucous. The adult leaves are dull, blue-green to bluish, sometimes glaucous, 60 to 130 mm long by 14 to 34 mm wide. The buds and fruits are often glaucous and occur in umbels of seven to thirteen. The buds have a long, horn-shaped operculum. The fruits are urn-shaped and 5 to 9 mm long and wide. The flowers are pale yellow, occurring mostly in spring.

DISTRIBUTION & HABITAT : Widespread in the western part of the State in the Great Victoria Desert and the northern ranges, from about the Gawler Ranges westwards and northwards. It occurs in scattered stands, but where it occurs it is usually the most common eucalypt. It grows on red sands, often on slight limestone rises, or on low sand dunes over limestone. Associated species include *E. gypsophila, E. eremicola* and *E. pimpiniana* in the Great Victoria Desert or *E. oxymitra* in the ranges in the far north-west of the State. *E. eucentrica* extends into W.A. and N.T.

CULTIVATION & USES : Occasionally cultivated, glaucous forms with their yellow flowers are quite ornamental. It grows in any soil that is well drained but it requires a hot dry climate. It is suited for use as a windbreak and for erosion control and for amenity planting.

NOTES : This is the northern form of *E. socialis*, and the two species are closely related and difficult to distinguish in places such as on northern Eyre Peninsula. *E. eucentrica* differs from *E. socialis* in its yellow (rather than cream) flowers and the larger leaves, buds and fruits, although leaf, bud and fruit dimensions overlap with the two species. *E. wyolensis* is the neotonous (remaining juvenile) form of *E. eucentrica* (likewise, *E. gillii* is the neotonous form of *E. socialis*), and the two species intergrade.

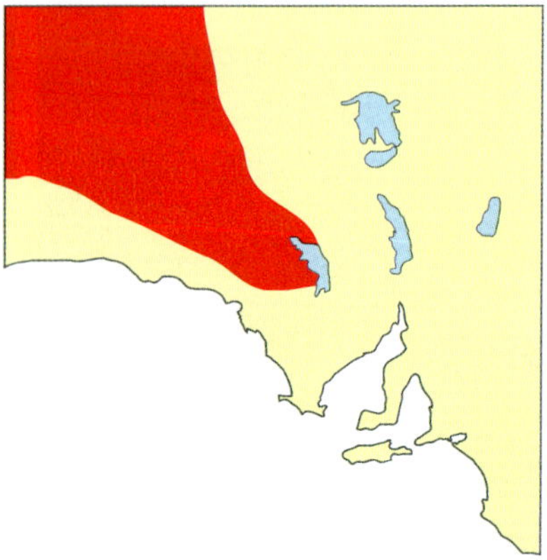

E. eucentrica on the Mimili to Fregon road, north-west of the Everard Ranges, S.A.

Eucalyptus wyolensis Boomsma

Wyola Lake Mallee

Genus equivalent: *Symphyomyrtus*; section: *Bisectaria*; series: *Subulatae*

NAME : Both the scientific and common names refer to Wyola Lake, the nearest geographical landmark to the population of this species.

DESCRIPTION : A mallee of scraggy habit, 4 to 6 metres tall. The bark is rough on the lower half of the stems, loose, grey, then smooth above. The leaves on the mature mallee are in the juvenile phase, being rounded, sessile (stem-clasping), grey and glaucous. They are 35 to 85 mm long by 30 to 75 mm wide. The buds and fruits are also glaucous. The operculum of the bud is horn-shaped. The flowers are pale yellow.

DISTRIBUTION & HABITAT : Known only from a small area north of Cook on the Ooldea Range in an area of about 4 x 2 km, just to the east of the Cook to Vokes Hill track in the southern Great Victoria Desert (about 10 km south of Wyola Lake). It grows on a plain of red sand with *E. concinna, E. canescens* subsp. *beadellii* and *E. eucentrica*. Endemic to South Australia.

CULTIVATION & USES : Only recently brought into cultivation, it is relatively slow growing, but flowers at a very young age. It requires a hot, sunny position in a well drained soil. It may prove superior to *E. gillii* in hotter parts of the State, being very ornamental with its grey foliage and yellow flowers. The foliage and flower buds could be used in floral arrangements.

NOTES : The large, sessile, rounded, grey leaves of this species make it very distinctive. It is similar to *E. gillii*, differing in its much larger leaves, buds and fruits, its yellow flowers and its western distribution. It is most closely related to *E. eucentrica*, differing in its more scraggy habit and the sessile, rounded, consistently grey and glaucous leaves. Intergrades between *E. wyolensis* and *E. eucentrica* are common west of the distribution of *E. wyolensis*, occurring almost to the Western Australian border, and have leaves on short petioles that are longer and more pointed than leaves of *E. wyolensis*.

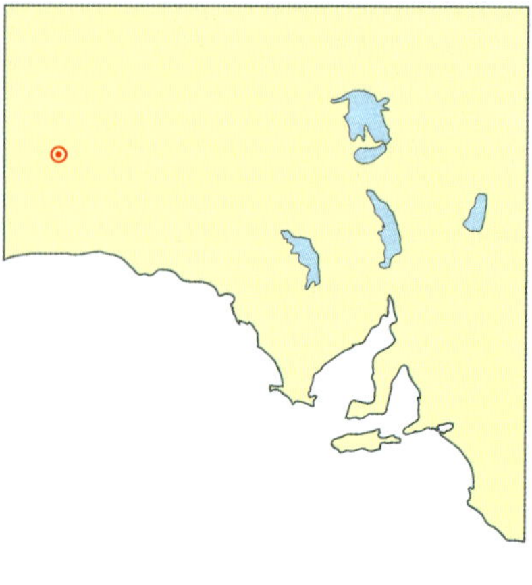

E. wyolensis on the Oak Valley to Tjuntjunjara track, south of Wyola Lake, Great Victoria Desert, S.A.

Eucalyptus gillii Maiden

Curly Mallee; Arkaroola Mallee; Silver Mallee

Genus equivalent: *Symphyomyrtus*; section: *Bisectaria*; series: *Subulatae*

NAME : Named in honour of Walter W. Gill (1851 - 1929), a Conservator of Forests for South Australia who sent the original specimens of *E. gillii* to J.H. Maiden.

DESCRIPTION : A mallee 3 to 6 metres tall, often of a scraggy habit. The bark is rough at the base or up to about a metre, fibrous and grey, then smooth above, light grey to orange. The twigs are glaucous. The juvenile leaves are short, sessile, dull, blue-grey and glaucous. The crown of a mature plant is composed mostly or entirely of juvenile leaves. Adult leaves are rare and are slightly longer and have petioles instead of being sessile. Buds and fruits are strongly glaucous, but otherwise very similar to *E. socialis* in size and shape. The flowers are cream to pale yellow.

DISTRIBUTION & HABITAT : Occurs in the northern Flinders Ranges in rocky areas on the slopes and ridges and less often on flat areas near the hills. It commonly grows in shallow loams over limestone outcrops. It is common in the rugged hills around Arkaroola but extends north to Mount Fitton and south to Bibliando. Plants intermediate between *E. gillii* and *E. socialis* occur as far south as near Yunta. It also occurs in the Barrier Range near Broken Hill in N.S.W.

CULTIVATION & USES : It has become common in cultivation in recent years, grown mainly as an ornamental because of its silvery leaves and creamy-yellow flowers that are often prolific on cultivated plants. It grows well in Adelaide, but grows much more vigorously in drier and hotter areas of the State. It may become scraggy with age but this can be rectified by heavy pruning.

NOTES : Closely related to *E. socialis*, differing mainly in the predominance of juvenile leaves on a mature mallee. Intergrades between the two are common on the ditributional margin of *E. gillii*. It is also similar to *E. wyolensis,* differing in the smaller leaves, buds and fruits; the paler flowers and the eastern distribution.

E. gillii on the Angepena to Balcanoona road, Gammon Ranges National Park, northern Flinders Ranges, S.A.

Eucalyptus socialis F. Muell. ex Miq.

Red Mallee; Summer Red Mallee

Genus equivalent: *Symphyomyrtus*; section: *Bisectaria*; series: *Subulatae*

NAME : From the Latin *socialis* meaning 'friendly' referring to it often being associated with other eucalypt species.

DESCRIPTION : A mallee, sometimes of poor form, 2 to 8 metres tall. The bark is usually rough and grey-brown over the lower stems, or occasionally completely smooth and cream to pinkish grey. The buds and twigs are sometimes glaucous. The adult leaves are dull to glossy and green to grey-green. The buds and fruits are in groups of seven to eleven. The fruits have fragile needle-like valves. The flowers are cream to pale creamy yellow.

DISTRIBUTION & HABITAT : Widespread from Eyre Peninsula east through the Murray Mallee into Victoria and New South Wales, including the Flinders Ranges, with a few outlying populations to the north as far as north-west of Oodnadatta. Also occurs in the Pilbara region of Western Australia and with isolated occurrences in central Queensland. It usually grows in mallee scrub on sands to heavy loams usually over shallow limestone. One of the most common and widespread mallees in South Australia.

CULTIVATION & USES : Often cultivated because of its ability to grow on a range of sites in dry areas of the State, including shallow limestone. It has a moderate growth rate and flowers at an early age. Some forms can be ornamental and it is suitable for broadscale planting.

NOTES : A variable species, with some forms that may become new species or subspecies in the future (eg. a narrow-leaved form in the northern Flinders Ranges and a glossy, green-leaved form widespread in S.A.), however, they are all included in *E. socialis* here. *E. socialis* differs from *E. eucentrica* in the smaller leaves, buds and fruits that are lightly or non-glaucous and the cream flowers, but the two species intergrade on northern Eyre Peninsula. Differs from *E. yalatensis* in the longer operculum; larger buds and fruits and the usually taller habit. Differs from *E. yumbarrana* in the smaller, thinner and often bluish adult leaves and much smaller buds and fruits.

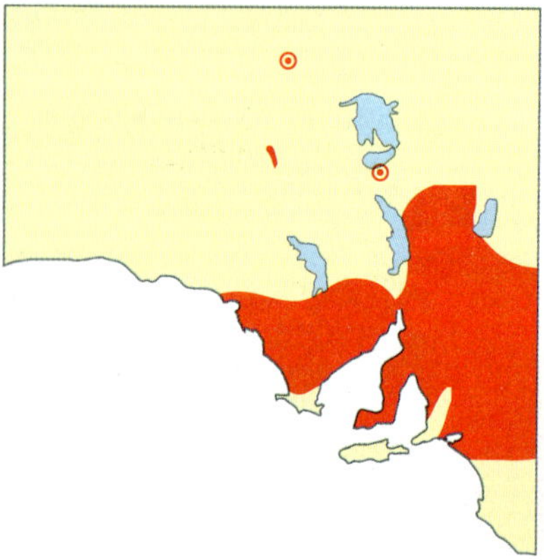

E. socialis on Corunna South, north of Iron Knob, looking south-west towards Lake Gilles, S.A.

Eucalyptus yalatensis Boomsma

Yalata Mallee

Genus equivalent: *Symphyomyrtus*; section: *Bisectaria*; series: *Subulatae*

NAME : Both the scientific and common names refer to Yalata Community on the Nullarbor Plain of South Australia.

DESCRIPTION : A low, spreading mallee to 4 metres tall and of equal or greater spread. The bark is usually rough and grey on most or some of the stems, or mostly smooth, especially on central Eyre Peninsula. All parts are non-glaucous. The adult leaves are dull and blue-green. The small buds and fruits occur in groups of seven to eleven. The buds have a sharply conical operculum. Fruits are cupular to hemispherical with fragile exserted valves that are broken off with age. The flowers are white.

DISTRIBUTION & HABITAT : Of mostly coastal distribution, from Balladonia in Western Australia to about Hincks Conservation Park on Eyre Peninsula in South Australia. It is the dominant eucalypt of the mallee vegetation along the Nullarbor cliffs. It is also scattered east of Tailem Bend in the southern Murray Mallee. It grows in sand or loam over limestone in mallee scrub, associated species include *E. oleosa* subsp. 'Coastal' and *E. gracilis* on the Nullarbor cliffs or *E. gracilis, E. diversifolia* and *E. phenax* on Eyre Peninsula and the Murray Mallee areas.

CULTIVATION & USES : Occasionally cultivated, it is very useful as a low windbreak in coastal sites or on difficult limestone soils. It flowers profusely as a small plant and would be suitable as a shrub for gardens and for broadscale planting, especially in difficult dry, coastal sites. It requires a well-drained soil. Plants have a moderate growth rate.

NOTES : Distinguished by the combination of low spreading habit; dull, bluish leaves; small buds with a conical operculum and the small delicate fruits.

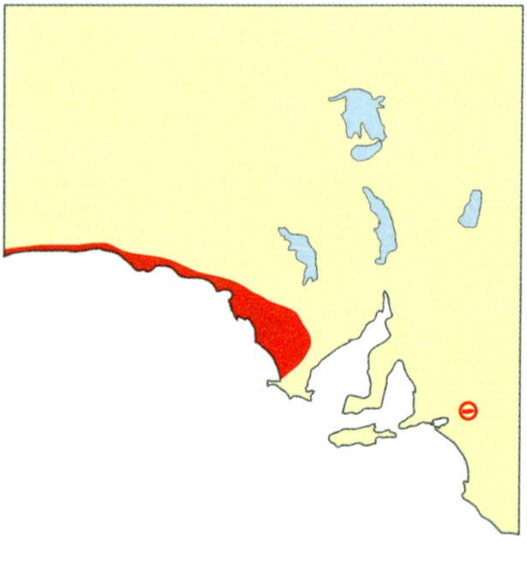

E. yalatensis east of Border Village near the W.A./S.A. border, Nullarbor Plain, S.A.

Eucalyptus peninsularis Nicolle ined.

Cummins mallee

Genus equivalent: *Symphyomyrtus*; section: *Bisectaria*; series: *Subulatae*

NAME : From the Latin *peninsula* (a narrow body of land) referring to its occurrence on Eyre Peninsula.

DESCRIPTION : A mallee, 4 to 7 metres tall. The bark is completely smooth and light grey to tan. The juvenile leaves are sessile and strongly decurrent (the leaf margins joining the stem lower down than the central vein). Adult leaves are very glossy and green. The buds and fruits are usually pendulous and occur in umbels of seven to eleven. The buds have a beaked operculum that is wider than the hypanthium (base) at the join. The fruits have a prominent constricting neck making them urn-shaped. The flowers are pale creamy-yellow.

DISTRIBUTION & HABITAT : Restricted to lower central Eyre Peninsula, being common in the Cummins to Yeelanna area but extending south and east towards Port Lincoln and the Koppio Hills. It occurs on flat or rarely hilly areas on loams to clay-loams associated with other mallee species such as *E. pileata, E. oleosa* and *E. calycogona* var. *spaffordii*. Endemic to South Australia.

CULTIVATION & USES : Unknown in cultivation. The glossy, green leaves and urn-shaped buds and fruits are quite attractive and it should be useful for mass plantings on roadsides and farms for screening, shelter and erosion control.

NOTES : This species was previously included in *E. flocktoniae*, a species restricted to south-west Western Australia, differing, among other things, in its less urn-shaped, erect buds and fruits. *E. peninsularis* can be distinguished from all other South Australian eucalypts by its strongly decurrent juvenile leaves. It may be confused with some glossy-leaved forms of *E. socialis*, however, the consistently smooth bark and the usually pendulous, prominently urn-shaped buds and fruits of *E. peninsularis* distinguish it. *E. peninsularis* is most closely related to *E. urna*, a non-lignotuberous tree (rather than a mallee), endemic to Western Australia.

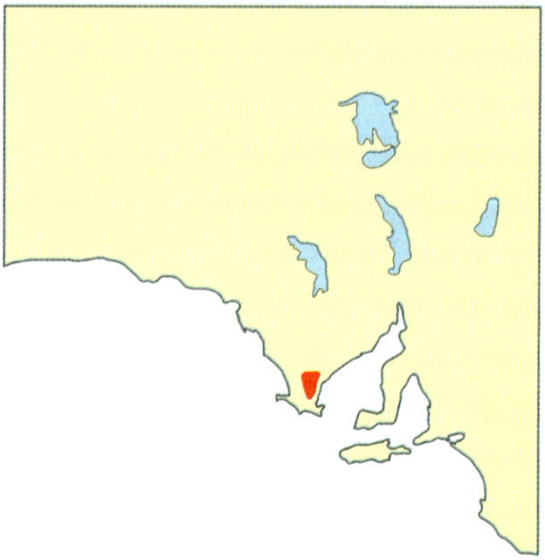

E. peninsularis on the Cummins to Yallunda Flat road, Eyre Peninsula, S.A.

Eucalyptus oxymitra Blakely

Sharp-capped Mallee

Genus equivalent: *Symphyomyrtus*; section: *Bisectaria*; series: *Curviptera*

NAME : From the Greek *oxys* meaning 'sharp' and *mitra* meaning 'head-dress', referring to the pointed operculum.

DESCRIPTION : A densely crowned mallee, 2 to 6 metres tall. The bark is usually smooth throughout, but in large mallees some loose, rough bark is present on the lower stems. The leaves are thick, dull and blue-green to grey. The buds and fruits occur in umbels of seven. Buds have a short, pointed operculum and the fruits are large, rounded and have sharp, exserted valves when dry. The branchlets, leaves, buds and fruits are often glaucous. The flowers are cream with yellow anthers.

DISTRIBUTION & HABITAT : Occurs in the area around the Mann, Musgrave, Everard and Birksgate Ranges in north-western South Australia and adjacent areas of Western Australia and Northern Territory. It occurs on small rocky rises or on the slopes and at the base of hills. It also grows on red sand dunes and the intervening plains and extends into the northern Great Victoria Desert. Associated eucalypts include *E. gamophylla, E. eucentrica, E. youngiana* and *E. mannensis*.

CULTIVATION & USES : Rarely cultivated. Plants are slow growing, especially in the south of the State. It would be worth trying in the northern half of the State where better results may be obtained. It would require a well drained soil and a hot, sunny position. Its low, dense habit makes it ideal for use as a windbreak and for erosion control in such areas.

NOTES : The mallee habit; the large, thick leaves and the moderately large, round and often glaucous fruit with protruding valves make this species easy to identify. *E. glomerosa* has similar leaves and fruits, but always has thick, rough bark on the lower stems, is never glaucous and has a round rather than sharply pointed operculum. The two species grow together in the northern Great Victoria Desert, with *E. glomerosa* occurring on the sides of sand dunes and *E. oxymitra* occurring in the swales between the dunes.

E. oxymitra near Currie Creek, west of Fregon, northern Great Victoria Desert, S.A.

Eucalyptus kingsmillii (Maiden) Maiden & Blakely subsp. *alatissima* Brooker & Hopper

Wing-fruited Mallee

Genus equivalent: *Symphyomyrtus*; section: *Bisectaria*; series: *Curviptera*

NAMES : The specific name honours W. Kingsmill (1865 - 1935), a former Member of the Legislative Council in Western Australia. The subspecific name from the Latin *alatus* meaning 'winged' and *issimus* meaning 'very much so', referring to the winged buds and fruits.

DESCRIPTION : A robust mallee 4 to 8 metres tall. The bark is rough and dark grey on the lower half of the stems, becoming ribbony then smooth thereafter. The branchlet, buds and immature fruits are glaucous. The adult leaves are thick, dull, and light green to blue-green. The pendulous (hanging) buds and fruits occur in umbels of three and are prominently winged. The pendulous flowers are pink to red with yellow anthers, usually occurring in autumn.

DISTRIBUTION & HABITAT : Scattered in the Great Victoria Desert from west of Vokes Hill to the Serpentine Lakes in South Australia and continuing west to at least Neale Junction in Western Australia. It grows on red sand on the plains between sand dunes. Associated species include *E. youngiana, E. concinna, E. glomerosa* and *E gongylocarpa.*

CULTIVATION & USES : Only recently brought into cultivation, so far it has proved slow growing and unreliable in Adelaide but it is faster growing in more arid areas. In hot, dry conditions it is highly suitable for gardens and parks because of its large, pendulous red to pink flowers and large winged buds and fruits which are highly ornamental and attractive to birds. It requires a well drained soil. Plants flower when quite small.

NOTES : A very ornamental and distinctive eucalypt, not likely to be confused with any other in South Australia. It differs from the related *E. youngiana* in the long peduncles and pedicels causing the flowers to be pendulous; the glaucous twigs and the smaller, glaucous buds and fruits. *E. kingsmillii* subsp. *kingsmillii*, an endemic of Western Australia, differs in having yellow flowers and non-glaucous twigs, buds and fruits.

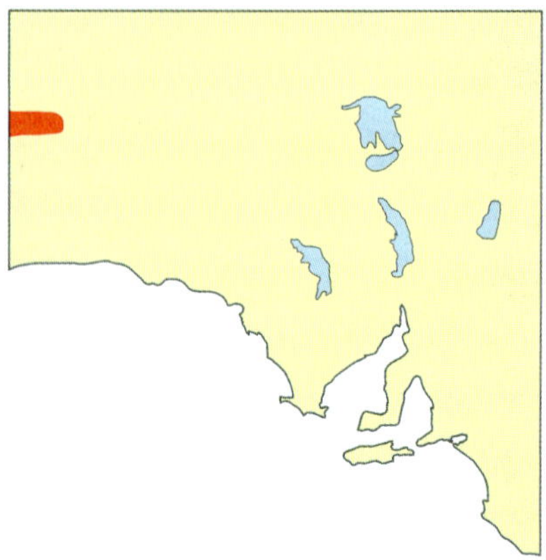

E. kingsmillii subsp. *alatissima* near Wanna Lakes on Anne Beadell Highway, Great Victoria Desert, W.A.

Eucalyptus youngiana F. Muell.

Ooldea Mallee; Large-fruited Mallee

Genus equivalent: *Symphyomyrtus*; section: *Bisectaria*; series: *Curviptera*

NAME : Named after Jess Young, a member of the Giles Exploratory party and collector of the type specimen.

DESCRIPTION : A large, heavily-crowned mallee, 2 to 8 metres tall. The bark is rough for up to 2 metres, thick and ribbony, then smooth and ribbony above, grey to cream. Adult leaves are thick, dull and light green to more commonly blue-green. The buds and fruits are in threes, being massive and broadly ribbed, sometimes weighing down the branches. Old fruits are often present on the ground under the mallees. The flowers are pale yellow to yellow or red, occurring mostly in winter and spring.

DISTRIBUTION & HABITAT : A common mallee of the Great Victoria Desert of South and Western Australia, occurring from south-east of Tarcoola and near Kondoolka Station to the Birksgate Range area and westwards. It grows on deep red sand on sand dunes and the intervening plains associated with a variety of other desert eucalypts. *E. youngiana* also grows on the rocky slopes and summit of Mount Finke, south-west of Tarcoola with *E. trivalvis*, however, it is not known to grow in rocky situations elsewhere.

CULTIVATION & USES : A commonly cultivated mallee for ornamental purposes, producing an abundance of large flowers in cultivation under good conditions, attracting nectar feeding birds. It is suited to large gardens and parks. It requires a well-drained soil in a sunny position and prefers hot, dry areas. Old plants can be rejuvenated by pruning back to ground level.

NOTES : A most spectacular mallee with the largest fruits in the genus. (*E. macrocarpa*, often claimed to have the largest fruits, has lighter but broader fruits than *E. youngiana*.) The large buds and fruits and colourful flowers make this mallee distinctive and not likely to be confused with any other South Australian eucalypt. Red flowers are more common in the western part of its distribution (eg. near Queen Victoria Spring, W.A.), while in South Australia the flowers are more often yellow.

E. youngiana on Mount Finke, looking south-west over the Great Victoria Desert, S.A.

E. youngiana on the Oak Valley to Tjuntjuntjara track, near Forrest Lakes, Great Victoria Desert, W.A.

Eucalyptus leptopoda Benth. subsp. *elevata*

L. Johnson & K. Hill

Whipstick Mallee

Genus equivalent: *Symphyomyrtus*; section: *Bisectaria*; series: *Curviptera*

NAMES : The specific name is from the Greek *leptos* meaning 'slender' and *pous* meaning 'foot', referring to the slender peduncles and pedicels. The subspecific name is from the Latin *elevatus* meaning 'raised', referring to the raised fruit disc of the subspecies compared to subspp. *leptopoda* and *subluta* (subsp. *arctata* also has a raised disc).

DESCRIPTION : A many-stemmed whipstick mallee with a rounded crown, 2 to 3 metres tall. The bark is smooth, pink to grey or cream. The adult leaves are dull and light green to blue-green. The buds and fruits occur in umbels of seven. The buds have a pointed operculum. The fruits are saucer-shaped, on slender peduncles and pedicels. The flowers are cream.

DISTRIBUTION & HABITAT : Known from a single clump of mallees in South Australia at 28° 31' 57"S 129° 56' 27"E , east of the Serpentine Lakes in the central Great Victoria Desert, but other populations probably exist nearby in such a remote area of similar habitat. It becomes increasingly common westwards in Western Australia, where it becomes frequent in the Neale Junction area of the Great Victoria Desert. It extends further west, in disjunct populations, almost as far as Geraldton. In the Great Victoria Desert it grows on red sand between dunes with *E. youngiana*.

CULTIVATION & USES : Rarely cultivated. All four subspecies of *E. leptopoda* are very attractive mallees with a lot of potential for ornamental use. Because of its small size it is suitable for home gardens and general landscaping use. Trials have shown it to have a moderate growth rate in Adelaide and to flower at an early age. It develops into an attractive plant in most soils provided they are well drained.

NOTES : This delightful mallee is unlikely to be confused with any other South Australian eucalypt. The other three subspecies of *E. leptopoda* occur only in the wheatbelt and goldfield regions of Western Australia.

E. leptopoda subsp. *elevata* north-west of Wanna Lakes on Anne Beadell Highway, Great Victoria Desert, W.A.

Eucalyptus glomerosa Brooker & Hopper

Jinjulu

Genus equivalent: *Symphyomyrtus*; section: *Bisectaria*; series: *Curviptera*

NAME : From the Latin *glomerosus meaning* 'globular' referring to the spherical buds.

DESCRIPTION : A mallee 4 to 7 metres tall. The bark is rough for 1 to 3 metres, thick, loose, flaky and grey to brown, then smooth above, grey to salmon-pink or coppery. The leaves are dull and blue-green. The new growth may be glaucous. Buds are in umbels of seven, globular in shape, on long pedicels. The fruits are rounded due to the broad convex disc, with exserted valves. The flowers are cream.

DISTRIBUTION & HABITAT : Widespread and common in the Great Victoria Desert of Western and South Australia, from north of Wyola Lake to south of the Mann Range and east to just north of Emu. It is restricted to red sand, preferring the crests and sides of sand dunes but also occurring on the interdunal plains. Commonly associated species include *E. gongylocarpa, E. eucentrica* and *E. youngiana.*

CULTIVATION & USES : Virtually unknown in cultivation, trials in Adelaide show it to be difficult to establish. More success may be had in hotter, drier parts of the State. It requires a very well drained soil in full sun. In such conditions it may be suited for windbreaks.

NOTES : Previously included with *E. ewartiana* (a native to W.A. only); *E. glomerosa* differs in its non-minniritchi type bark (*E. ewartiana* has minniritchi bark like that of *E. orbifolia*) and the more slender pedicels. The globular buds and fruits distinguish *E. glomerosa* from other South Australian mallees, especially *E. oxymitra*, which has larger buds and fruits and a pointed operculum, unlike *E. glomerosa.*

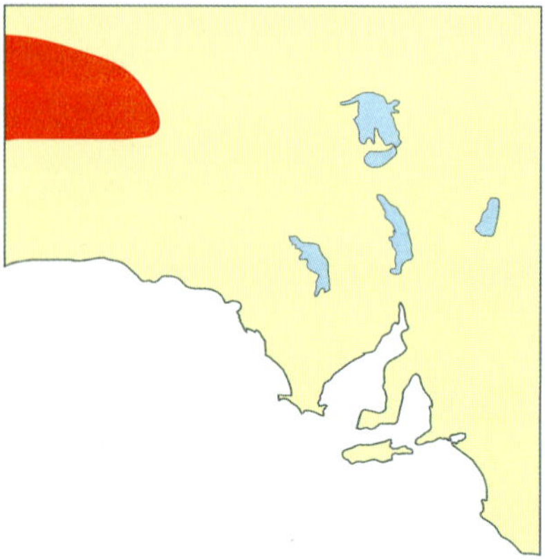

E. glomerosa near Turner Hill, northern Great Victoria Desert, S.A.

Eucalyptus orbifolia F. Muell.

Round-leaved Mallee

Genus equivalent: *Symphyomyrtus*; section: *Bisectaria*; series: *Orbifoliae*

NAME : From the Latin *orbis* meaning 'round' and *folium* meaning 'leaf', referring to the rounded adult leaves.

DESCRIPTION : A spreading or sprawling mallee, with a dense or open crown, 1 to 3 metres tall. The bark is minniritchi type (red-brown older bark, pealing off in short, thin, longitudinal strips that curl to reveal greenish-grey bark beneath). The twigs are glaucous. The adult leaves are rounded to heart-shaped, dull, blue-grey and glaucous when new. The buds and fruits occur in umbels of seven and are strongly glaucous. The operculum (cap) is 4 to 6 mm long and the fruits 8 to 13 mm wide. The flowers are pale yellow.

DISTRIBUTION & HABITAT : Restricted to a few isolated populations in the Musgrave Ranges in South Australia, where it grows on the upper slopes and summit of the higher mountains, such as Mount Woodroffe and Mount Everard. It grows in coarse sand between sandstone boulders on steep slopes, sometimes associated with *E. trivalvis* and *C. eremaea*. It is also known from the Petermann Ranges in Western and central Australia and is more common in the Macdonnell Ranges of Northern Territory. The type form of *E. orbifolia* comes from the goldfields of Western Australia, where it has been erroneously called *E. lata*.

CULTIVATION & USES : A very ornamental plant occasionally cultivated in Adelaide. It prefers hot, dry conditions, and will adapt to most well drained soils. It should be cultivated more often in gardens and parks in hotter areas of the State and would make an attractive large container plant. Plants are generally slow growing but flower when quite small.

NOTES : The only eucalypt native to South Australia with minniritchi type bark. True *E. orbifolia* is endemic to the goldfields of Western Australia. The central ranges form, described here, differs in the narrower leaves; shorter rounded operculum (7 to 11 mm long) and smaller fruits than the goldfields form, and may represent a distinct taxon. Also related to *E. websteriana* (restricted to W.A.) differing in the strongly glaucous twigs, buds and fruit; the broader leaves and the larger buds (operculum 2 to 5 mm long in *E. websteriana*) and fruit.

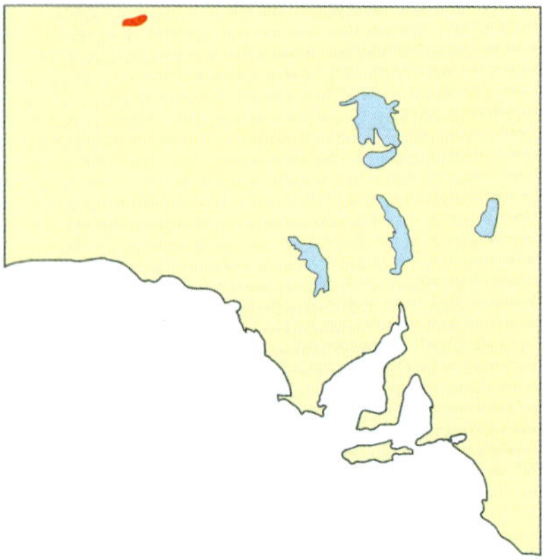

E. orbifolia on Mount Hermannsberg, Macdonnell Ranges, N.T.

Eucalyptus leptophylla F. Muell. ex Miq.

Narrow-leaved Red Mallee; Slender-leaved Mallee

Genus equivalent: *Symphyomyrtus*; section: *Bisectaria*; series: *Porantherae*

NAME : From the Greek *leptos* meaning 'narrow' and *phyllon* meaning 'leaf', referring to the narrow adult leaves.

DESCRIPTION : A mallee, often thin-stemmed, 2 to 6 metres tall. The bark is usually smooth and cream and grey to coppery. The adult leaves are narrow, very glossy and green, with numerous conspicuous oil glands when held up to the light. The buds and fruits occur in umbels of seven to thirteen. The buds are usually prolific and conspicuous, becoming bright orange to red near flowering. The fruits have a light-coloured, level disc. The flowers are white, with all stamens having anthers.

DISTRIBUTION & HABITAT : Widespread from about Balladonia in Western Australia, eastwards through the mallee regions of South Australia into Victoria and New South Wales. Also scattered throughout the southern half of the Great Victoria Desert. It grows on sandy soils on plains in mallee scrub, but also occurs in scattered populations in the Mount Lofty Ranges and on Kangaroo Island. It is often associated with a number of species such as *E. gracilis, E. oleosa, E. socialis, E. cyanophylla* and *E. diversifolia.*

CULTIVATION & USES : A delightful mallee and very ornamental because of its smooth slender stems, narrow leaves and plentiful coloured buds. It should be cultivated more often in gardens, parks and along roadsides. Plants have a moderate growth rate. The buds and foliage could be used in the cut flower trade.

NOTES : Easily confused with *E. gracilis*, differing in the usually smooth bark; the colourful buds before flowering; the broad white disc to the fruit and all stamens possessing anthers. The leaf venation is also different (more parallel in *E. gracilis*). Easily distinguished from most other mallee species with which it may grow by its glossy, green, narrow leaves and small but plentiful buds and fruit. Mallees in the Great Victoria Desert are not typical in possessing some rough bark and may represent a distinct taxon.

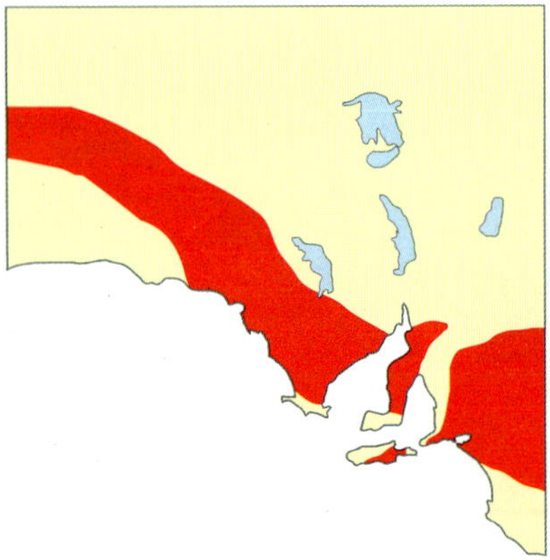

E. leptophylla north of Koonibba, Yumbarra Conservation Park, S.A.

Eucalyptus gracilis F. Muell.

Yorrell; White Mallee

Genus equivalent: *Symphyomyrtus*; section: *Bisectaria*; series: *Heterostemones*

NAME : From the Latin *gracilis* meaning 'slender', referring to the sometimes graceful and slender habit of this species.

DESCRIPTION : A mallee 4 to 7 metres tall, or a small tree 6 to 12 metres tall on Boston Island. The bark is rough on the lower stems, weakly tessellated, grey to brown, then smooth, olive green or tan to cream above, sometimes completely smooth throughout. Adult leaves are glossy and green, with many oil glands when held up to the light. Buds and fruits are in umbels of seven. The small fruits have a thin rim and enclosed valves. The flowers are white, the outer filaments lacking anthers and often twisted. Flowers occur in autumn, winter and spring.

DISTRIBUTION & HABITAT : Widely distributed in southern Australia, from the Norseman area of Western Australia to western Victoria and south-western New South Wales, including most areas of southern South Australia including the southern Great Victoria Desert. Absent from the Nullarbor Plain, most of Kangaroo Island, the wetter parts of the Mount Lofty Ranges, the northern Flinders Ranges and the lower South-east. It grows on deep red sands to clay-loams, usually over limestone. Commonly associated species include *E. oleosa, E. porosa, E. phenax* and *E. brachycalyx.*

CULTIVATION & USES : Often cultivated for broadscale plantings such as on farms and roadsides. Plants have a moderate growth rate. It is especially suitable for limestone soils in the drier, southern parts of the State.

NOTES : One of the most frequently encountered mallees of the wheatbelt areas of South Australia. It can be easily confused with *E. leptophylla*, but differs in the more robust form; the often rough bark; the more acute side venation; the flowers with outer filaments lacking anthers and the more barrel-shaped fruits with a thin rim. Closely related to *E. calycogona* differing in the smaller, rounded rather than square (in cross section) buds and fruits.

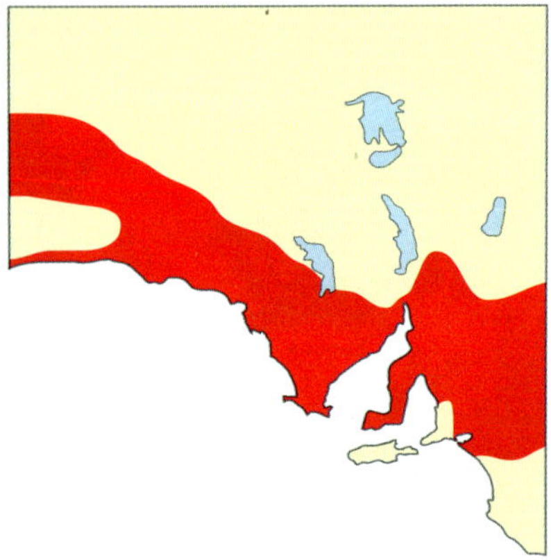

E. gracilis near Pelican Lagoon on the Penneshaw to Kingscote road, Kangaroo Island, S.A.

Eucalyptus calycogona Turcz. subsp. *calycogona*

Square-fruited Mallee; Gooseberry Mallee; Red Mallee

Genus equivalent: *Symphyomyrtus*; section: *Bisectaria*; series: *Heterostemones*

NAME : From the Greek *calyx* (the calyx) and *gonia* meaning 'angle', referring to the angled buds and fruits.

DESCRIPTION : A mallee, stems usually erect, 3 to 5 metres tall. The bark is completely smooth, cream to grey. The adult leaves are 40 to 90 mm long by 6 to 13 mm wide, glossy and green, with many oil glands when held up to light. The buds and fruits are much longer than wide and have four longitudinal ribs, causing the fruit to be square in cross section. The fruits are 6 to 14 mm long by 3 to 6 mm wide and have a constricting neck near the rim. The flowers occur in winter to spring and are white to dark pink, with the outer stamens lacking anthers.

DISTRIBUTION & HABITAT : Occurs on Eyre Peninsula, where it is fairly common, preferring heavier soils, sometimes over limestone, in mallee scrub with a variety of other eucalypts such as *E. oleosa, E. phenax, E. leptophylla* and *E. peninsularis*. There are also a few small populations in mallee scrub between Currency Creek and Strathalbyn, far disjunct from the main Eyre Peninsula populations. It is also widespread in the wheatbelt and southern goldfield regions of Western Australia.

CULTIVATION & USES : The pink flowering forms are very ornamental and bird attracting when in flower, one such cultivar known as *E. calycogona* 'Jubilee'. It has a moderate growth rate and prefers heavy soils in areas of rainfall below about 650 mm annually. It is suited to mass plantings on roadsides and large gardens. Flower colour cannot be guaranteed, even if the seed source is known.

NOTES : The pink flowering forms are most common on eastern Eyre Peninsula. It is similar to *E. calycogona* subsp. 'Eastern', differing in the smaller, more erect habit; the smooth bark and the smaller adult leaves, buds and fruits. *E. calycogona* subsp. *calycogona* is also similar to var. *staffordii*, with which it may grow, differing in the smaller, thinner adult leaves and the much smaller, more weakly ribbed buds and fruits.

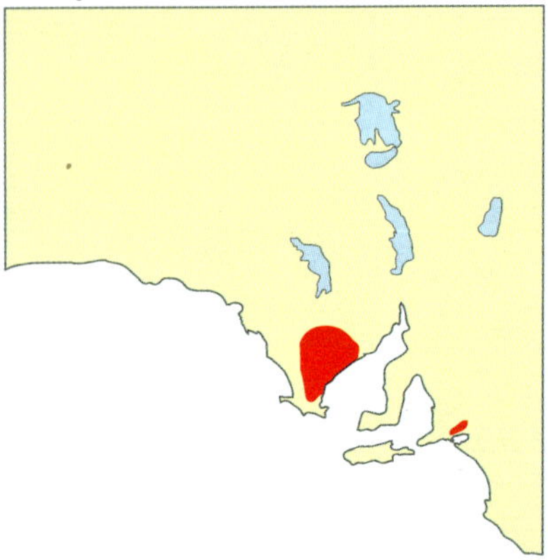

E. calycogona subsp. *calycogona* on the Kimba to Cleve road, east of Darke Peak, Eyre Peninsula, S.A.

Eucalyptus calycogona Turcz. subsp. 'Eastern'

Square-fruited Mallee; Gooseberry Mallee; Red Mallee

Genus equivalent: *Symphyomyrtus*; section: *Bisectaria*; series: *Heterostemones*

NAME : An undescribed subspecies occupying the eastern part of the distribution of *E. calycogona*.

DESCRIPTION : A mallee, usually of spreading habit, 2 to 6 metres tall. The bark is rough for 1 to 2 metres, finely fissured and slightly tessellated (like that of *E. gracilis*), grey, then smooth above, cream to olive-grey or orange. The adult leaves are 45 to 100 mm long by 8 to 15 mm wide, glossy and green, with many oil glands when held up to the light. The buds and fruits are similar to *E. calycogona* subsp. *calycogona* but marginally larger (buds 10 to 12 mm long by 3.5 to 5.5 mm wide and fruits 8 to 13 mm long by 4 to 7 mm wide). The flowers are white to rarely pale pink, occurring in winter to spring, with the outer stamens lacking anthers.

DISTRIBUTION & HABITAT : Occurs eastwards of Spencer Gulf, through the Murray Mallee and upper South-east into adjacent Victoria and just into New South Wales. It occurs in heavy soils or in low well-drained areas in mallee scrub, associated with species such as *E. oleosa, E. phenax, E. socialis and E. dumosa.*

CULTIVATION & USES : Less often seen in cultivation than *E. calycogona* subsp. *calycogona.* Although the flowers are slightly larger, they are not known to be the dark pink like sometimes seen in subsp. *calycogona.* It seems to be a more vigorous and spreading mallee than subsp. *calycogona* and may be more suitable for windbreaks. Again it prefers loams and clays.

NOTES : This is the eastern form of *E. calycogona* and differs from subsp. *calycogona* and var. *staffordii* in the rough bark on the lower stems. The leaves, buds and fruits are larger than those of subsp. *calycogona,* but still much smaller than those of var. *staffordii.*

E. calycogona subsp. 'Eastern' south of Pinnaroo on the Bordertown road, S.A.

Eucalyptus calycogona Turcz. var. *staffordii* Blakely

Yeelanna Mallee

Genus equivalent: *Symphyomyrtus*; section; *Bisectaria*; series: *Heterostemones*

NAME : Named after W.J. Spafford (1884 -1962), collector of the type specimen. Blakely misspelt the name in the original description.

DESCRIPTION : A mallee 4 to 6 metres tall. The bark is completely smooth, cream to grey. The adult leaves are 75 to 110 mm long by 13 to 24 mm wide, thick, glossy and green, with many oil glands when held up to light. The buds and fruits have four prominent longitudinal ribs or wings, making them square in cross section. The buds are 12 to 15 mm long by 4 to 7 mm wide and the fruits are 11 to 13 mm long by 6 to 9 mm wide (larger than subspp. *calycogona* and 'Eastern'). The flowers, only recorded as white in colour, occur in winter to spring with the outer stamens lacking anthers.

DISTRIBUTION & HABITAT : Restricted to lower central Eyre Peninsula, mainly around Yeelanna and south to Cummins. It grows on loams in mallee scrub with species such as *E. calycogona* subsp. *calycogona, E. peninsularis, E. diversifolia* and *E. leptophylla*. Endemic to South Australia.

CULTIVATION & USES : Virtually unknown in cultivation. The large four winged buds and fruits are attractive. It would be suitable for general landscape use and broadscale planting. Its environmental requirements should be the same as that of *E. calycogona* subsp. *calycogona*.

NOTES : Although this variety sometimes grows in association with subsp. *calycogona*, the two do not intergrade, indicating this variety would be best treated as a separate species (where the spelling would be corrected as *E.* 'spaffordii'). It differs from *E. calycogona* subsp. *calycogona* in its larger, thicker adult leaves and the larger, more prominently ribbed buds and fruits. From subsp. 'Eastern' it differs also in the completely smooth bark. *E. calycogona* var. *staffordii* is a species of restricted distribution that in the past has often been overlooked by botanists.

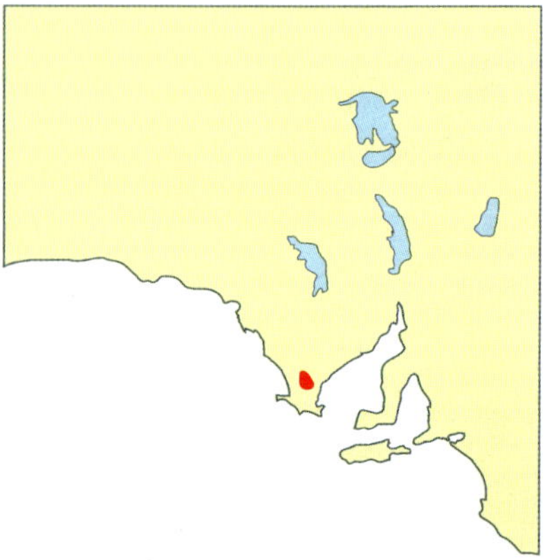

E. calycogona var. *spaffordii* on the Yeelanna to Cummins road, Eyre Peninsula, S.A.

Eucalyptus pimpiniana Maiden

Pimpin Mallee

Genus equivalent: *Symphyomyrtus*; section: *Dumaria*; series: unnamed

NAME : Both the scientific and common names are from the Aboriginal name for this plant.

DESCRIPTION : A low, bushy mallee, 0.5 to 2 metres tall. The bark is smooth and white, tan and pink. The branchlets possess pith glands. The adult leaves are thick, dull and blue-grey, with numerous oil glands obscuring the venation pattern. The pendulous buds and fruits occur in umbels of seven to fifteen, and are often faintly longitudinally ribbed. The flowers are bright yellow.

DISTRIBUTION & HABITAT : Known mainly from the Ooldea Range in the southern Great Victoria Desert in South Australia, from Barton to the southern arm of Unnamed Conservation Park, extending into Western Australia in very scattered population west as far as Lake Minigwal in the south-western Great Victoria Desert where it is again common. It grows on sandplains or dunes of coarse red sand with species such as *E. concinna, E. eucentrica* and *E. yumbarrana*.

CULTIVATION & USES : The low habit, the large, greyish leaves and masses of bright yellow flowers makes this mallee highly ornamental and bird attracting. It is occasionally cultivated, and develops into an attractive small shrub, although it may become straggly if not heavily pruned every 5 or 6 years. Plants are generally slow growing but flower heavily when quite small. It would make a good ornamental plant for small or large gardens and parks in the drier parts of the State. It requires a very well-drained soil in a hot, sunny position.

NOTES : A distinctive eucalypt with no close relatives. Its low habit; thick grey leaves with hidden venation due to numerous oil glands; yellow flowers and long, cylindrical buds and fruits readily distinguish this species from any other.

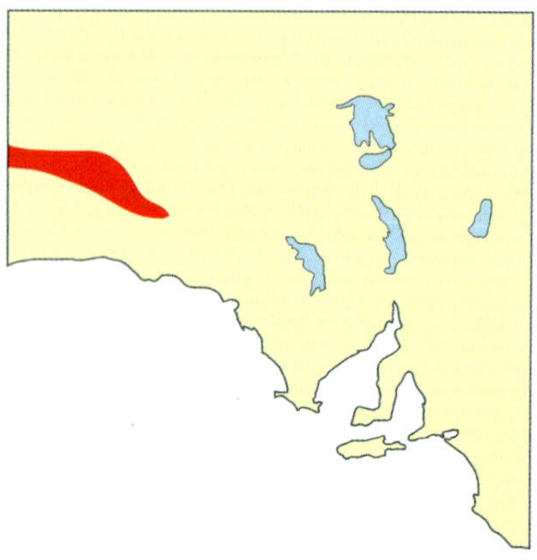

E. pimpiniana on the Watson to Maralinga road, Ooldea Range, Great Victoria Desert, S.A.

E. pimpiniana south of Lake Minigwal, Great Victoria Desert, W.A.

Eucalyptus incrassata Labill. var. *incrassata*

Yellow Mallee; Ridge-fruited Mallee; Lerp Mallee

Genus equivalent: *Symphyomyrtus*; section: *Dumaria*; series: *Tetrapterae*

NAME : From the Latin *incrassatus* meaning 'thickened' referring to the thick leaves of this species.

DESCRIPTION : A large, heavily crowned mallee 4 to 8 metres tall. The bark is rough for a metre or two, loose, ribbony and grey-brown, then smooth, cream to grey-brown, shedding in ribbons that often remain in the branches. The adult leaves are thick and firm, glossy and green, with many oil glands when held up to light. The buds and fruits are in umbels of three to more commonly seven. The buds have a beaked operculum. The cylindrical fruits are smooth to ribbed. The flowers are white, or occasionally turn pink before the stamens fall.

DISTRIBUTION & HABITAT : Occurs throughout most of Eyre Peninsula, eastwards to Yorke Peninsula, the Murray Mallee and the upper South-east. It prefers sandy soils and often grows only on the low sandy ridges present in parts of the wheatbelt, being absent from the heavier soils between the ridges. It occurs in mallee scrub with species such as *E. socialis, E. leptophylla, E. arenacea and E. phenax*. It extends into adjoining areas of Victoria and New South Wales and also occurs disjunctly in coastal Western Australia from south of Madura westwards.

CULTIVATION & USES : Occasionally cultivated, it is suited to windbreaks and soil erosion control in sandy soils. Plants are generally quick growing and have very few pests and diseases. It generally become straggly and unattractive with age but this can be rectified with heavy pruning. The flowers are somewhat showy and attract nectar feeding birds.

NOTES : *E. incrassata* var. *incrassata* is closely related to var. *angulosa*, differing in the narrower juvenile and adult leaves; the smaller, less ribbed buds and fruits and the fruits that are more often in sevens The two varieties intergrade and are difficult to distinguish near Waitpinga. Also similar to *E. ceratocorys*, differing in the narrower adult leaves; the smaller, less ribbed buds and fruits and the fruits with a less prominent neck.

E. incrassata var. *incrassata* near Tooligie on the Lock to Cummins road, Eyre Peninsula, S.A.

Eucalyptus incrassata Labill. var. *angulosa* Schauer

Ridge-fruited Mallee

Genus equivalent: *Symphyomyrtus*; section: *Dumaria*; series: *Tetrapterae*

NAME : From the Latin *angulosus* meaning 'strongly angled' referring to the branchlets, buds and fruits.

DESCRIPTION : A mallee, often with a dense, heavy crown, 2 to 5 metres tall. The bark is smooth but very ribbony and light grey to brown (the ribbons often hanging in the branches). The adult leaves are thick, glossy and dark green. The buds and fruits are ribbed and occur in umbels of three. The buds have a beaked operculum. The ribbed fruits are often abundant and being heavy, may weigh down the branches. The flowers are white, sometimes turning pink before falling.

DISTRIBUTION & HABITAT : Occurs on the southern tip of Eyre Peninsula south of Port Lincoln, such as in Lincoln National Park and around Sleaford Mere, with a small occurence at Newland Head near Waitpinga. It grows close to the coast in shallow loams over limestone in dense mallee scrub with species such as *E. diversifolia*. The true (type) form of *E. incrassata* var. *angulosa* is endemic to South Australia. Mallees currently called *E. incrassata* var. *angulosa* on the south coast of Western Australia differ from the type form in many respects and will be described as a new species in the future. Although some reference books state either or both subspecies of *E. incrassata* to grow on Kangaroo Island, the two species have never been recorded there.

CULTIVATION & USES : Occasionally cultivated, it forms a dense mallee or shrub excellent for windbreaks and screening in difficult coastal areas. It will tolerate salty soils providing they are well drained and it withstands very strong winds. Plants have a reasonably quick growth rate. The large flowers and fruits are ornamental and attract nectar feeding birds.

NOTES : The coarse, glossy green leaves and the large buds and fruits in threes are distinctive. It differs from var. *incrassata* in the broader, thicker leaves and the much larger, more strongly ribbed buds and fruits consistently in threes. On Eyre Peninsula the two species do not occur together and are easily separated. However, at Newland Head the two grow together and intergrade completely.

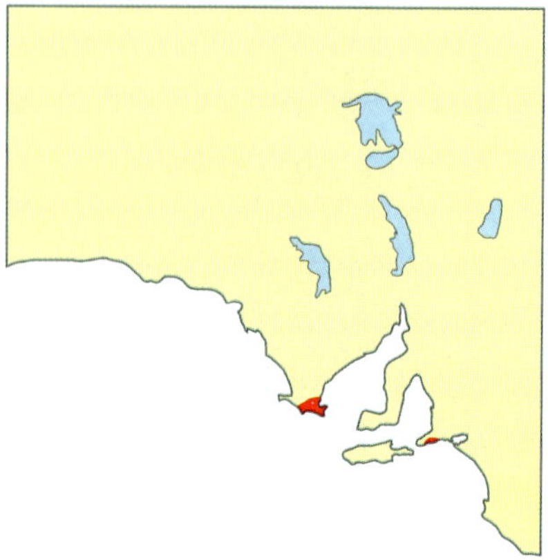

E. incrassata var. *angulosa*
south of Port Lincoln, Eyre
Peninsula, S.A.

Eucalyptus ceratocorys (Blakely) L. Johnson & K. Hill

Horn-capped Mallee

Genus equivalent: *Symphyomyrtus*; section: *Dumaria*; series: *Tetrapterae*

NAME : From the Greek *ceras* meaning 'horn' and *corys* meaning 'helmet', referring to the horn-shaped operculum.

DESCRIPTION : A straggly mallee 3 to 6 metres tall. The bark is rough for 1 to 2 metres, loose and ribbony, then smooth, ribbony and tan to grey. Adult leaves are glossy and dark green with many oil glands visible when held up to the light. The ribbed buds and fruits are in groups of seven and are pendulous. The buds and especially the fruits have a constricted neck 2/3 the way up giving the fruit an urn-shape. The flowers are pale yellow to cream.

DISTRIBUTION & HABITAT : Once thought to be restricted to Western Australia, it is now known to be common in Yumbarra Conservation Park in the far southern Great Victoria Desert of South Australia from the eastern edge of Yumbarra Conservation Park to about Micherie Rockhole. It grows on red sand on and between the dunes associated with species such as *E. yumbarrana* and *E. concinna*. Also occurs on the south-western edge of the Great Victoria Desert and parts of the northern wheatbelt in Western Australia.

CULTIVATION & USES : This desert mallee is rarely cultivated but is quick growing in sands and loams in the Adelaide area and areas of lower rainfall (< 500 mm annually). It often forms a straggly growth habit but this can be avoided by lopping it back to near ground level every 6-10 years. The large buds and fruits, which are often prolific, are bird attracting and along with the thick, glossy green leaves are ornamental.

NOTES : Closely related to *E. incrassata*, it differs from var. *incrassata* in the broader, thicker leaves and the larger, strongly ribbed buds and fruits that are constricted to form a neck. From var. *angulosa* it differs in the non-coastal distribution; the buds and fruits consistently in sevens; the longer peduncles and pedicels making the buds and fruits consistantly pendulous and the constricted buds and fruits.

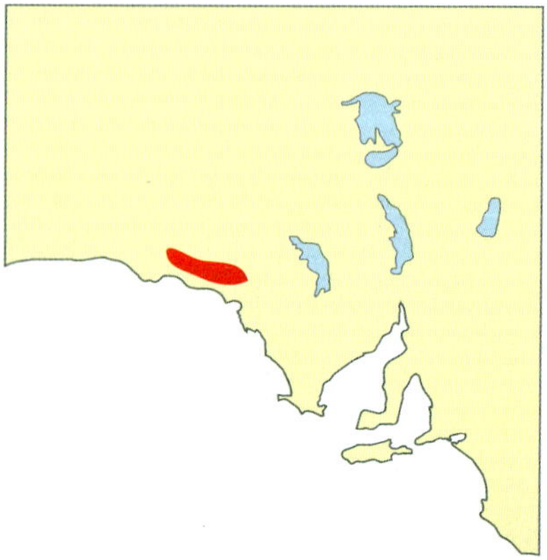

E. ceratocorys south of Lake Minigwal, Great Victoria Desert, W.A.

Eucalyptus rugosa R. Br. ex Blakely

Kingscote Mallee

Genus equivalent: *Symphyomyrtus*; section: *Dumaria*; series: *Torquatae*

NAME : From the Latin *rugosus* meaning 'wrinkled', referring to the buds and fruits.

DESCRIPTION : Varies from a low dense mallee (on the mainland) to a large erect mallee (on Kangaroo Island). On exposed coastal cliffs on Eyre Peninsula it is forced to grow as a ground cover under 50 cm tall. The Bark is smooth and cream or grey to yellow-green or salmon-pink. Pith glands are present in the branchlets. The leaves are thick, glossy and green. The buds and fruits are in umbels of seven to eleven. Buds are usually coarsely ribbed and the operculum (cap) is narrower than the hypanthium (base) where they join. The fruits are also usually ribbed or angled. Seeds are grey. The flowers are white, occurring mostly in summer.

DISTRIBUTION & HABITAT : Of coastal distribution, from the Coorong to western Eyre Peninsula and common on southern Kangaroo Island and lower Yorke Peninsula. It commonly grows in shallow loams over limestone in dense mallee scrub. Associated species include *E. albopurpurea, E. gracilis, E. diversifolia* and *E. phenax*. Endemic to South Australia.

CULTIVATION & USES : Occasionally cultivated, it is very useful for harsh coastal sites, including limestone soils, where it can be used for windbreaks and screening. It is moderately ornamental and has a moderate growth rate.

NOTES : This species and *E. brachycalyx* intergrade in the Coonalpyn area. *E. rugosa* has a more coastal distribution than *E. brachycalyx* and differs in the consistently smooth bark; the thicker, broader leaves and the larger, coarser buds and fruits. *E. rugosa* may also be confused with *E. phenax*, but can be distinguished by the grey seed; the operculum always being narrower than the hypanthium at the join and the seven to eleven-flowered buds and fruits.

E. rugosa near Fowl Bay, Yorke Peninsula, S.A.

Eucalyptus brachycalyx Blakely

Gilja; Chindoo Mallee

Genus equivalent: *Symphyomyrtus*; section: *Dumaria*; series: *Torquatae*

NAME : From the Greek *brachys* meaning 'short' and *calyx* (the calyx or hypanthium), referring to short hypanthium (bottom half of the flower bud).

DESCRIPTION : A mallee 4 to 7 metres tall. The bark is rough for up to two metres, grey, then smooth, grey to cream or occasionally completely smooth throughout. The branchlets possess pith glands. The adult leaves are glossy and green. The buds and fruits occur in umbels of seven to eleven. The buds have a rounded to flat-topped, ribbed operculum which is usually narrower than the hypanthium at the join. The fruits have exserted, spreading valves when dry. The flowers are white.

DISTRIBUTION & HABITAT : Of scattered occurence, from the Murray Mallee westwards to Eyre Peninsula and the far south-eastern Great Victoria Desert, disjunctly to the south coast of Western Australia. It is most common on loams over limestone on northern Eyre Peninsula and fringing areas of the Nullarbor Plain. North of Ceduna it occurs between sand dunes with *E. yumbarrana* and *E. ceratocorys*. A small isolated population of mallees most closely related to *E. brachycalyx* occurs in the River Torrens gorge east of Adelaide on a steep, rocky slope.

CULTIVATION & USES : Occasionally seen in cultivation. This mallee is not particularly ornamental, but is useful for mass plantings with other mallee species for roadside plantings and on farms for use as screening, windbreaks and wind erosion control. Plants have a moderate growth rate.

NOTES : This is the inland form of *E. rugosa*, and the two species intergrade in the Coonalpyn area. *E. brachycalyx* differs from *E. rugosa* in the often rough bark; the narrower, thinner leaves and the smaller buds and fruits. *E. brachycalyx* may be confused with *E. oleosa*, but can be distinguished by the presence of pith glands (absent in *E. oleosa*), and the spreading valves of the dried fruit.

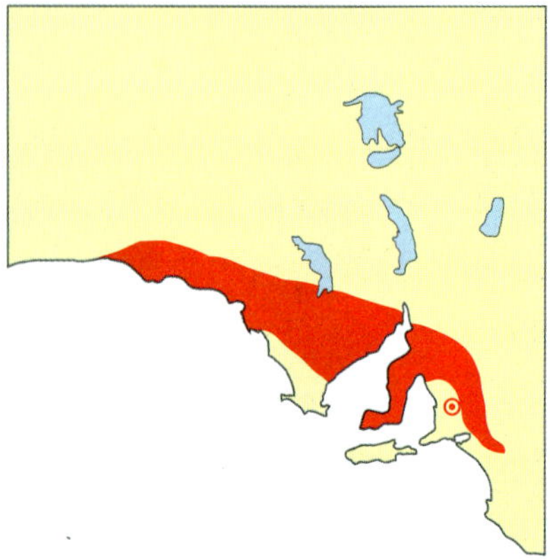

E. brachycalyx on Gorge Road above the River Torrens, Mount Lofty Ranges, S.A.

Eucalyptus concinna Maiden & Blakely

Great Victoria Desert Mallee

Genus equivalent: *Symphyomyrtus*; section: *Dumaria*; series: *Torquatae*

NAME : From the Latin *concinnus* meaning 'neat' or 'pretty', probably referring to the habit or overall appearance of the plant.

DESCRIPTION : A mallee, 3 to 6 metres tall. The bark is usually rough for about a metre, dark grey, then smooth, or completely smooth, cream to grey or tan. The adult leaves are extremely glossy and green. The buds and fruits occur in groups of seven to eleven. The operculum is rounded to flattened, lightly ribbed and usually wider than the hypanthium (base of the bud). The operculum sometimes becomes bright red before flowering. Bud and fruit size is quite variable over the species distribution. The flowers are white.

DISTRIBUTION & HABITAT : Widespread and common throughout the Great Victoria Desert of South Australia to south-east of Tarcoola and to Pinkawillinie Conservation Park and near Lake Macfarlane. However, it is at its best in the remote central Great Victoria Desert. It occurs west to around Kalgoorlie and Carnegie in Western Australia. It occurs on red sand on sandplains and the swales and sides of sand dunes. Commonly associated species include *E. youngiana, E. eucentrica E. leptophylla* and *E. glomerosa*.

CULTIVATION & USES : Little tried in cultivation, it has shown to have a moderate growth rate in the southern part of the State. It should be grown more for ornamental purposes, especially in hotter areas of the State, because of its attractive, glossy, green leaves and attractive buds that become reddish before flowering. It requires a very well drained soil and a hot, sunny position.

NOTES : This species has perhaps the glossiest leaves of all the eucalypts and the shining leaves makes it conspicuous in the field. It can be distinguished from other glossy-leaved species in the Great Victoria Desert by the flattened, ribbed operculum. Closely related to *E. brachycalyx*, differing in its more arid habitat; the broader, even more glossy leaves and the larger, 'fatter' buds and fruits.

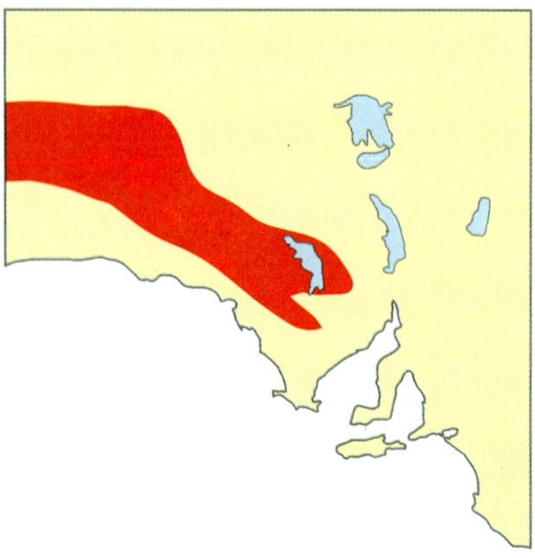

E. concinna south-east of Ooldea, on the edge of the Great Victoria Desert, S.A.

Eucalyptus pileata Blakely

Capped Mallee; Ravensthorpe Mallee (WA)

Genus equivalent: *Symphyomyrtus*; section: *Dumaria*; series: *Rufispermae*

NAME : From the Latin *pileatus* meaning 'cap-shaped', referring to the operculum.

DESCRIPTION : In South Australia it forms an erect mallee 4 to 7 metres tall. The bark is completely smooth and light grey to tan. The adult leaves are fairly thick (firm), glossy, and dark green. The buds and fruits occur in umbels of seven and have noticeable peduncles and pedicels. The bud operculum is ribbed and wider than the hypanthium (bud base) at the join, turning red before flowering. The cup-shaped fruits are smooth. The flowers are cream, occurring mostly in summer.

DISTRIBUTION & HABITAT : In South Australia known only from a very small area in the Koppio Hills on lower Eyre Peninsula near Yallunda Flat, and northwards to the Cummins area, where it is very scattered in woodland in hilly areas on the slopes and ridges with *E. odorata* and *E. petiolaris* or in flatter areas with *E. peninsularis*. *E. pileata* is much more common and widespread in the south-west of Western Australia.

CULTIVATION & USES : Rarely seen in cultivation, but should be cultivated more. In ornamental characteristics it is superior to related mallees in its more erect form; the glossy, green leaves and larger ribbed buds and large fruits. It would grow in any well drained soil in a sunny position.

NOTES : A rare species in South Australia, much more common in Western Australia. *E. pileata* differs from the related *E. phenax* and *E. dumosa* in the more erect habit; the glossy, dark green adult leaves; the larger buds with a ribbed operculum wider than the hypanthium at the join and the larger, more cupular fruits.

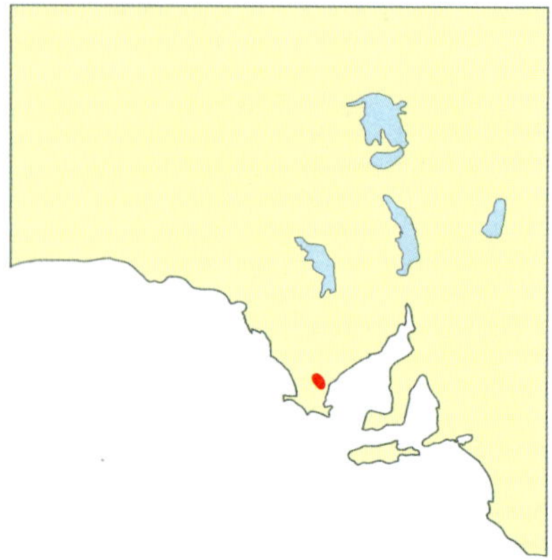

E. pileata south of Yallunda Flat, Koppio Hills, Eyre Peninsula, S.A.

Eucalyptus calcareana Boomsma

Nundroo Gum; Nundroo Mallee

Genus equivalent: *Symphyomyrtus*; section: *Dumaria*; series: *Rufispermae*

NAME : From the Latin *calcareus* meaning 'limey', referring to its occurrence on soils associated with limestone.

DESCRIPTION : A tree up to 15 metres tall, sometimes reduced to a robust mallee to 4 metres high. The bark is completely smooth, cream to grey or orange-tan, decorticating in long ribbons. Pith glands are present in the branchlets. The adult leaves are dull at first, slightly blue-green, becoming glossy and green with time. The buds and fruits occur in groups of seven to nine. The fruits are cupular to obconical. The flowers are white.

DISTRIBUTION & HABITAT : Of strictly coastal distribution, from north-western Eyre Peninsula westwards along the coast into Western Australia as far as south of Madura. It is most common on the edge of the Nullarbor Plain along the Eyre Highway where it often forms large, pure stands. It grows in shallow loams over limestone, sometimes with *E. oleosa* and *E. gracilis* present.

CULTIVATION & USES : In cultivation *E. calcareana* forms a small, spreading, densely crowned tree suitable for shade and windbreaks. It has the ability to grow in exposed sites in very shallow soils. Plants are relatively quick growing and flower at an early age.

NOTES : On northern and western Eyre Peninsula, *E. calcareana* becomes more mallee-like and becomes less coarse in the leaves, buds and fruits and appears to intergrade with *E. dumosa* and *E. phenax* subsp. *phenax* over large areas. However, *E. calcareana* is generally a coastal species and west of Eyre Peninsula it can generally be distinguished from these species in its more robust, often tree habit and in being generally coarser, having larger, thicker leaves and larger buds and fruits. The adult leaves are also dull and blue-green at first before turning glossy and green with age.

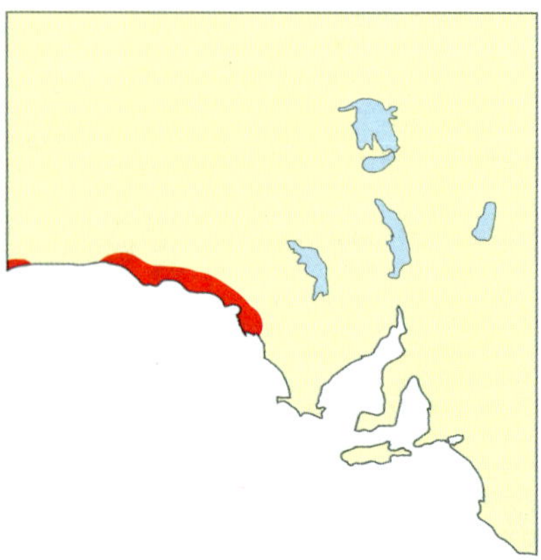

E. calcareana on the Colona to Ooldea track, Nullarbor Plain, S.A.

Eucalyptus conglobata (R. Br. ex Benth) Maiden

Port Lincoln Mallee; Cong Mallee

Genus equivalent: *Symphyomyrtus*; section: *Dumaria*; series: *Rufispermae*

NAME : From the Latin *conglobatus* meaning 'made like a ball', referring to the tightly clustered fruits.

DESCRIPTION : Usually a mallee 3 to 6 metres tall with a dense crown. On Boston Island, however, it develops into a slender tree to 14 metres tall with a terminal canopy. The bark is smooth, ribbony and cream to coppery, sometimes with a little ribbony-rough bark near the base. The branchlets possess pith glands. The leaves are thick, dull on new growth, quickly becoming glossy and green with age. The buds and fruits occur in tightly grouped clusters of seven, lack pedicels and have stubby or no peduncles. Individual fruits are wider than long and often flat sided due to crowding. The flowers are white. Seeds are glossy and ruby-red coloured.

DISTRIBUTION & HABITAT : Occurs in dense mallee scrub on the southern tip of Eyre Peninsula and on the adjacent Taylor and Boston Islands. The soils are fertile loams over limestone. It also occurs along the southern coast of Western Australia at least as far east as Toolinna Cove. Associated species include *E. gracilis, E. diversifolia* and *E. incrassata* var. *angulosa*.

CULTIVATION & USES : It is extremely useful for exposed, coastal sites including those with limestone present. Plants are relatively quick growing and suitable for windbreaks and erosion control. It will tolerate areas of high rainfall provided soils are well-drained.

NOTES : Closely related to *E. phenax*, especially subspecies 'Kangaroo Island'. *E. conglobata* differs in its larger buds and fruits that are wider than long; the shorter peduncles and lack of pedicels and the thicker, broader adult leaves.

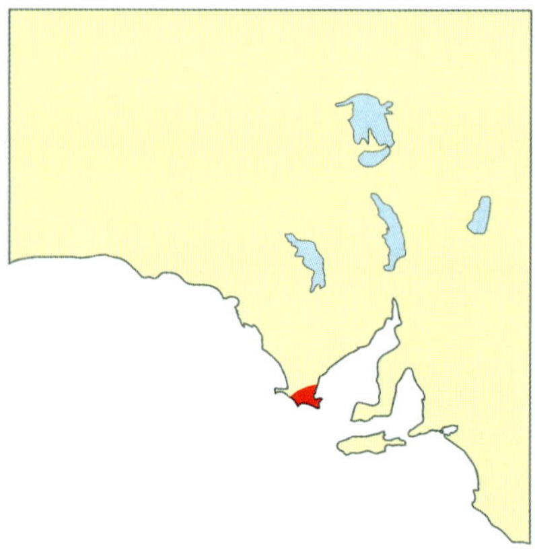

E. conglobata in Lincoln National Park, south of Port Lincoln, Eyre Peninsula, S.A

E. conglobata on Boston Island, east of Port Lincoln, S.A.

Eucalyptus phenax Brooker & Slee subsp. 'Kangaroo Island'

Kangaroo Island Mallee

Genus equivalent: *Symphyomyrtus*; section: *Dumaria*; series: *Rufispermae*

NAMES : An unnamed taxon related to *E. phenax* and *E. conglobata*, restricted mainly to Kangaroo Island.

DESCRIPTION : A mallee, often of erect form, 3 to 8 metres tall. The bark is usually smooth, light grey to tan and ribbony, although some ribbony-rough bark may be present near the base. The adult leaves are glossy and green. Buds and fruits are in umbels of seven. The peduncles are notably long (> 7 mm long) compared to subsp. *phenax* and *E. conglobata*. The fruits are about equal in length to width and occur in crowded clusters. The flowers are white. The seeds are glossy, red-brown.

DISTRIBUTION & HABITAT : Occurs sporadically on north-eastern Kangaroo Island and the adjacent mainland south coast near Waitpinga. It grows in mallee scrub usually on flat terrain in sands and loams although near Waitpinga it grows in hilly terrain in gravelly clay. Commonly associated species include *E. cneorifolia, E. odorata, E. rugosa* and *E. cosmophylla*. Endemic to South Australia.

CULTIVATION & USES : Not known in cultivation. It would require well-drained soils and is suited for coastal areas for windbreaks and shelter. Plants are relatively quick growing. It may be useful for broadscale planting such as along roadsides.

NOTES : This subspecies differs from *E. phenax* subsp. *phenax* in the broader, thicker adult leaves; the longer peduncles (to 7 mm long in subsp. *phenax*) and the broader, more crowded buds and fruits. The two subspecies intergrade north and east of Goolwa on the Mainland. It differs from *E. conglobata* in the much longer peduncles and smaller buds and fruits.

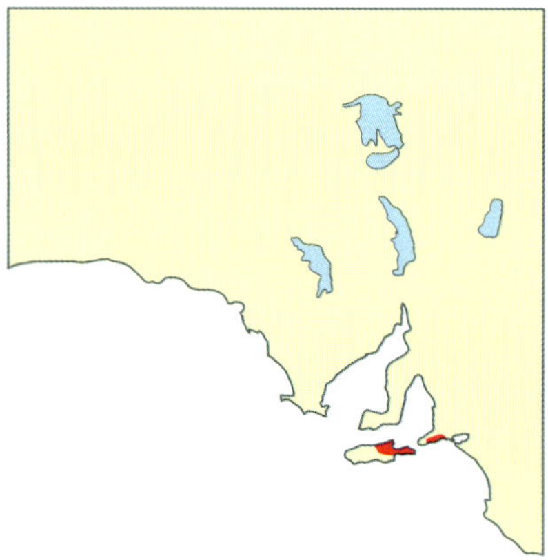

E. phenax subsp. 'Kangaroo Island' south-west of Kingscote, Kangaroo Island, S.A.

Eucalyptus phenax Brooker & Slee subsp. *phenax*

White Mallee

Genus equivalent: *Symphyomyrtus*; section: *Dumaria*; series: *Rufispermae*

NAME : From the Greek *phenax* meaning 'imposter' referring to the species past erroneous use of the name *anceps.*

DESCRIPTION : A mallee, 2 to 6 metres tall. The bark is mostly smooth and light grey to tan or coppery, sometimes with a little ribbony rough bark near the base. The adult leaves are glossy and green to dark green, however from a distance the mallees may appear slightly blue-green due to the duller new growth. The buds and fruits are in groups of seven. Buds are sessile or with pedicels to 3 mm long, with a ribbed operculum while the fruits are usually smooth. The flowers are white, usually occurring in summer. Seeds are glossy, red-brown.

DISTRIBUTION & HABITAT : Occurs from far western Victoria through the Murray Mallee to Yorke Peninsula and much of Eyre Peninsula. Also occurs in south-west Western Australia. In South Australia it is common in the southern Murray Mallee region. It grows in sands, loams or light clays, in mallee scrub with a variety of other mallees such as *E. socialis, E. oleosa, E. calycogona, E. odorata, E. gracilis* and *E. incrassata.*

CULTIVATION & USES : Occasionally cultivated, good forms produce masses of buds that become orange near flowering. Other than this it is not considered a highly ornamental mallee. Plants are fairly quick growing and suitable for windbreaks and wind erosion control.

NOTES : This species was previously incorrectly called *E. anceps*, a name now only correctly used as a synonym of *E. rugosa. E. phenax* can be confused with the closely related *E. dumosa; E. phenax* differs in the glossy, green adult leaves; shorter pedicels and peduncles and often slightly smaller buds and fruits. *E. phenax* subsp. *phenax* differs from *E. phenax* subsp. 'Kangaroo Island' in the usually narrower adult leaves; shorter peduncles; often longer pedicels and the less crowded fruit that are longer than wide (about equal in length and width in subsp. 'Kangaroo Island').

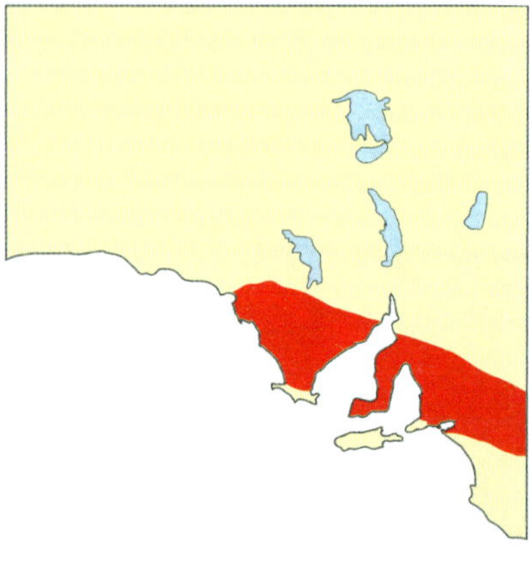

E. phenax subsp. *phenax* near Karoonda on the Murray Bridge to Loxton road, S.A.

Eucalyptus dumosa A. Cunn. ex Oxley

White Mallee; Dumosa Mallee; Congoo Mallee; Waikerie Mallee

Genus equivalent: *Symphyomyrtus*; section: *Dumaria*; series: *Rufispermae*

NAME : From the Latin *dumosus* meaning 'bushy', probably referring to its mallee habit.

DESCRIPTION : A mallee 3 to 6 metres tall. The bark is mostly smooth, white to pinkish grey. A little rough, thin, fibrous grey bark may be present near the base of large mallees. Pith glands are present in the branchlets. The adult leaves are dull and green to grey-green. The buds and fruits occur in umbels of seven to nine. The operculum is short and pointed. The fruits are cupular to barrel-shaped, occurring on individual pedicels (non-sessile). The flowers are white.

DISTRIBUTION & HABITAT : Widespread but scattered through the Murray Mallee, extending to the northern Flinders Ranges and onto northern Eyre Peninsula with some occurrences on the northern Adelaide Plains (as far south as Folland Park). It grows in mallee scrub on the plains with species such as *E. socialis, E. gracilis and E. leptophylla*. In the northern Flinders Ranges, it occurs on the slopes and ridges of rocky ranges. The Eyre Peninsula form has greener leaves than the eastern, typical form.

CULTIVATION & USES : Occasionally cultivated, this mallee has limited ornamental appeal but is useful for broadscale planting for roadside screening, windbreaks and erosion control. Plants have a moderate growth rate. It is suited to areas of rainfall between 300 and 650 mm annually. It will grow in any soil but prefers heavier soils (clays and loams).

NOTES : A variable species with many close relatives with which it can be easily confused. It differs from *E. cyanophylla* in the smaller adult leaves, buds and fruits and the usually greener leaves. *E. dumosa* can also be easily confused with *E. phenax* subsp. *phenax*, but differs in its dull, often bluish-green leaves, (only the new growth is dull in *E. phenax*) and the buds and fruits with pedicels (individual stalks). The Eyre Peninsula form, with its greener leaves, may represent a separate taxon.

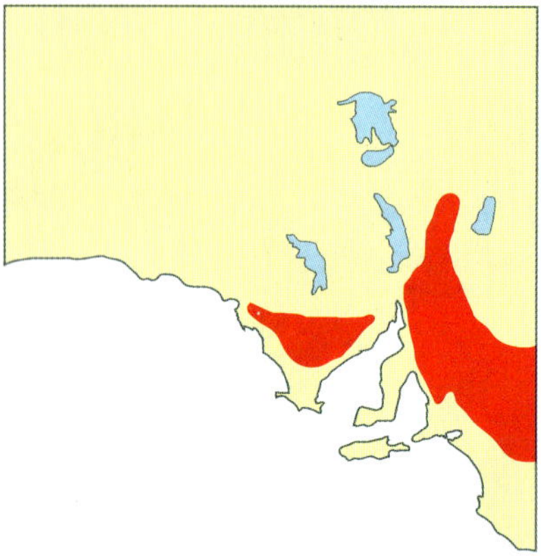

E. dumosa north of Pinnaroo towards Paruna, S.A.

Eucalyptus percostata Brooker & Lang

Rib-capped Mallee

Genus equivalent: *Symphyomyrtus*; section: *Dumaria*; series: *Rufispermae*

NAME : From the Latin *per* meaning 'very' and *costatus* meaning 'ribbed', referring to the strongly ribbed operculum of the bud.

DESCRIPTION : A mallee, 2 to 5 metres tall. The bark is smooth and cream and grey to coppery or pink. The adult leaves are dull, light green to blue-grey and sometimes slightly glaucous. The buds and fruits occur in umbels of seven. The buds have a strongly ribbed operculum (cap) that is clearly wider than the hypanthium (bud base) where they join. The operculum usually turns red before flowering. The fruits are smooth or ribbed. The flowers are white, occurring mainly in winter. Seeds are glossy, red-brown.

DISTRIBUTION & HABITAT : A rare mallee, known from a few sites in the southern Flinders Ranges from Devils Peak in the north to about The Bluff east of Port Pirie in the south, where it occupies the slopes and footslopes of hills. It grows in clay-loams in open mallee scrub or low woodland. It can be seen most prominently on the walking track to the summit of Devils Peak. Commonly associated species include *E. odorata*, *E. cladocalyx* and *E. socialis*. Endemic to South Australia.

CULTIVATION & USES : Almost unknown in cultivation, it would require the same climatic conditions as *E. dumosa*, but is suited to areas of heavier soil or slightly higher rainfall. Plants have a moderate growth rate.

NOTES : Closely related to *E. dumosa* differing mainly in the strongly ribbed operculum that is wider than the hypanthium at the join. *E. dumosa* grows to the north of *E. percostata* in the Flinders Ranges. It is also similar to *E. pileata*, differing mainly in the dull, often bluish adult leaves and the consistently many-stemmed mallee habit. A form of *E. percostata* with large smooth fruits occurs as small remote populations on the Mawson Plateau in the northern Flinders Ranges.

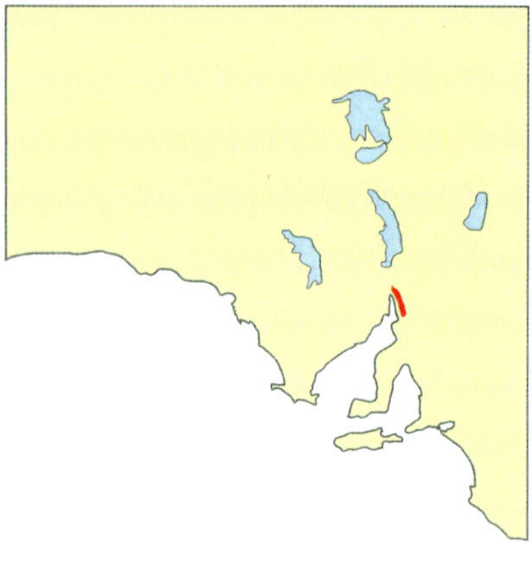

E. percostata at the base of Devils Peak, near Quorn, southern Flinders Ranges, S.A.

Eucalyptus cyanophylla Brooker

Blue-leaved Mallee; Murraylands Mallee

Genus equivalent: *Symphyomyrtus*; section: *Dumaria*; series: *Rufispermae*

NAME : From the Greek *cyaneos* meaning 'dark blue' and *phyllon* meaning 'leaf', referring to the bluish adult leaves.

DESCRIPTION : A mallee 2 to 6 metres tall. The bark is mostly smooth and ribbony, light grey to tan, usually with a short stocking of loose rough bark at the base. The twigs, buds and fruits are never glaucous. The adult leaves are dull, blue-grey and sometimes lightly glaucous. The buds and fruits occur in umbels of seven. The operculum of the bud is ribbed. The fruits are almost smooth to lightly ribbed. Flowers are white, occurring mainly in winter to early spring. The seeds are glossy, red-brown.

DISTRIBUTION & HABITAT : Restricted to the Murray Mallee region in South Australia and Victoria. In South Australia it is common south of the River Murray and is conspicuous in the Loxton area. It occurs in mallee scrub, in sandy soils, often associated with species such as *E. socialis, E. oleosa* and *E. leptophylla*.

CULTIVATION & USES : Occasionally cultivated, it will grow in light or heavy soils. It is suited to erosion control and windbreaks and is fairly ornamental with its bluish leaves and could be used for broadscale planting such as roadside planting. Plants have a moderate growth rate.

NOTES : Closely related to *E. dumosa* and *E. percostata*, differing mainly in the larger buds and fruits and the broader, thicker and consistently blue-grey adult leaves. Somewhat similar to *E. cretata,* differing in the non-glaucous twigs, buds and fruits and the often rough bark on the lower stems.

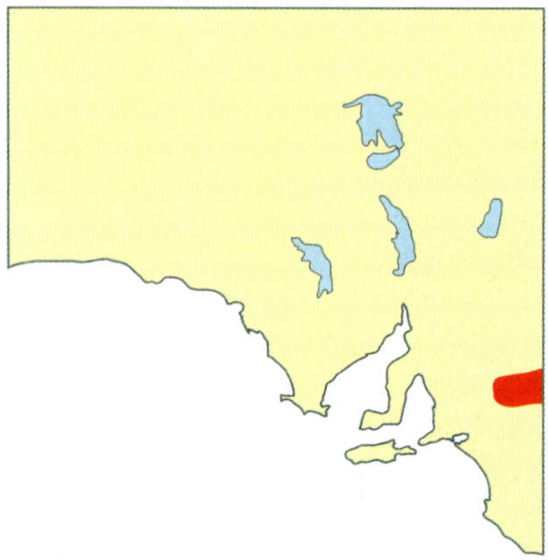

E. cyanophylla near Loxton, S.A.

Eucalyptus cretata Brooker & Lang

Darke Peak Mallee

Genus equivalent: *Symphyomyrtus*; section: *Dumaria*; series: *Rufispermae*

NAME : From the Latin *cretatus* meaning 'marked with chalk', referring to the white-glaucous twigs, buds and fruits.

DESCRIPTION : Usually a large mallee or small tree, to 8 metres tall, reduced to a straggly mallee on the rocky slopes of Carappee Hill. The bark is smooth, white and grey to coppery, shedding in ribbons. The twigs, buds and fruit are strongly glaucous. The adult leaves are dull, grey-green and glaucous but with age become greener and sometimes slightly glossy. Buds and fruits are in umbels of seven, the buds having a ribbed operculum usually wider than the hypanthium (bud base). The fruits are smooth to ribbed. The flowers are white, occurring in summer. Seeds are glossy, red-brown.

DISTRIBUTION & HABITAT : Scattered on central Eyre Peninsula from Caralue Bluff and Lock to near Cowell, being particularly common in the Darke Peak and Carappee Hill areas. It grows on plains and low hills in loams and clays. Common associated eucalypts include *E. calycogona* subsp. *calycogona* and *E. porosa.* Endemic to South Australia.

CULTIVATION & USES : Rarely cultivated, but has potential as an ornamental small tree or mallee for parks and gardens in areas of low to moderate rainfall. Trials show it to be adaptable to most soils including those with limestone. Plants have a moderate growth rate. The smooth, coloured bark and the glaucous, ribbed buds are very ornamental.

NOTES : Distinguished from other species on Eyre Peninsula by the strongly glaucous twigs, buds and fruits. Differs from the similar *E. canescens* and *E. gypsophila* (both from the Great Victoria Desert) in the smooth bark, shorter pedicels and usually more pointed operculum. Intergrades between *E. cretata* and *E. dumosa* are known on Eyre Peninsula and are distinguished by their slightly glaucous twigs.

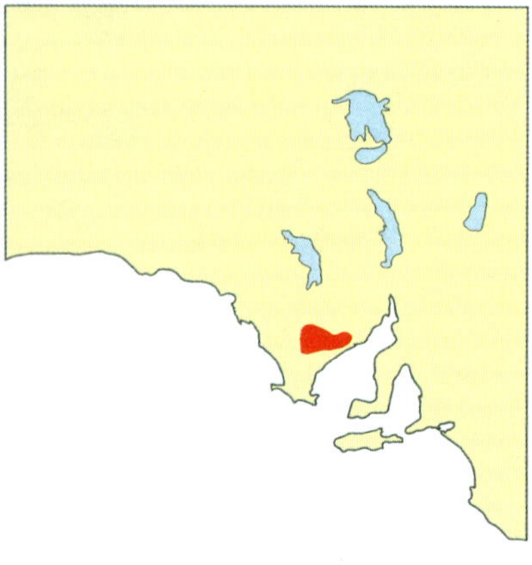

E. cretata near Darke Peak, Eyre Peninsula, S.A.

Eucalyptus canescens Nicolle subsp. *canescens*

Ooldea Range Mallee

Genus equivalent: *Symphyomyrtus*; section: *Dumaria*; series: *Rufispermae*

NAME : From the Latin *canescens* meaning 'becoming grey', referring to the grey general appearance of the plant caused by the greyish adult leaves and the glaucous twigs, buds and fruits.

DESCRIPTION : A mallee, usually dense in habit, 2 to 6 metres tall. The bark is rough, thick and flaky over most of the stems, grey to red-brown or yellow-brown. The twigs possess pith glands and are strongly glaucous. Adult leaves are ovate to broad-lanceolate, dull, blue-grey and glaucous. Buds and fruits glaucous, in umbels of seven. The buds have a ribbed, rounded operculum. The fruits are 10 to 16 mm long by 9 to 14 mm wide, ribbed, with 4 or 5 valves. The flowers are cream. Seeds are glossy, red-brown.

DISTRIBUTION & HABITAT : Known only from the Ooldea Range, a sand dune system extending from Ooldea to north of Cook in the southern Great Victoria Desert of South Australia. It is most easily seen on the Cook to Vokes Hill track. It grows in deep red sand on sand dunes and sand plains, associated species include *E. pimpiniana, E. eucentrica* and *E. concinna.* Endemic to South Australia

CULTIVATION & USES : Almost unknown in cultivation, it requires a well drained soil in hot areas. In clay soils it grows slowly and becomes lanky. This mallee is very ornamental and warrants growing in drier parts of the State (rainfall below 500 mm) where it would make an ornamental plant for gardens and parks. It may also be suitable for windbreaks and wind erosion control.

NOTES : Distinguished from related species by its consistently strongly glaucous twigs, buds and fruits; its large, broad adult leaves and the large, ribbed buds and fruits. *E. canescens* and *E. gypsophila* intergrade north of the Ooldea Range. From a distance *E. canescens* subsp. *canescens* looks similar to the relatively unrelated *E. eucentrica*, with which it often grows, however the two are readily distinguished by their buds and fruits. Distinguished from *E. cretata* especially in the rough, fibrous bark on the stems (completely smooth in *E. cretata*).

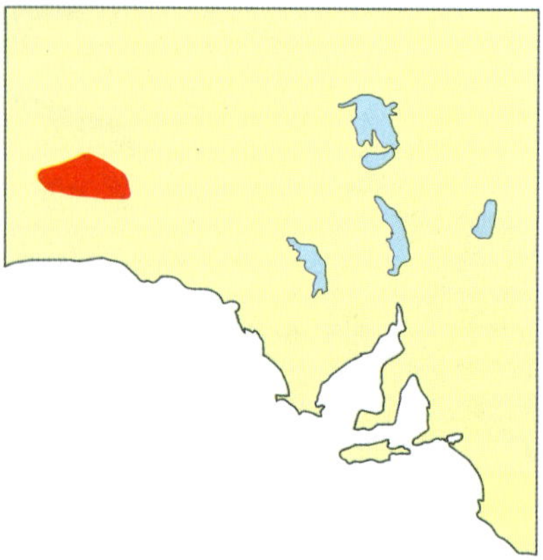

E. canescens subsp. *canescens* on the Cook to Vokes Hill track, Great Victoria Desert, S.A.

Eucalyptus canescens Nicolle subsp. *beadellii* Nicolle

Beadell's Mallee

Genus equivalent: *Symphyomyrtus*; section: *Dumaria*; series: *Rufispermae*

NAME : The name commemorates Len Beadell (1923 - 1995), who surveyed over 5000 km of roads to open up the Great Victoria, Gibson and Great Sandy Deserts to present day travellers.

DESCRIPTION : A mallee 2 to 4 metres tall. The bark is rough over the lower half of the stems, loose, flaky and grey-brown. The twigs possess pith glands and are non-glaucous. The adult leaves ovate to broad-lanceolate, dull to slightly glossy and green. The buds and fruits are non-glaucous, occurring in umbels of seven. The operculum is ribbed, rounded and clearly wider than the hypanthium (bud base) at the join. The fruits are 12 to 14 mm long by 11 to 13 mm wide, with 4 or 5 valves. The flowers are cream. Seeds are glossy, red-brown.

DISTRIBUTION & HABITAT : Of very restricted distribution known only from a sand plain north of Cook in the southern Great Victoria Desert, where it grows in deep red sand with *E. wyolensis, E. concinna* and *E. pimpiniana*. Endemic to South Australia.

CULTIVATION & USES : Unknown in cultivation. It would require light, sandy soils and would prefer the drier, hotter parts of the State where it would be suited for windbreaks and wind erosion control. The large club-shaped buds are very attractive.

NOTES : This rare mallee is similar to *E. canescens* subsp. *canescens* but is completely non-glaucous and has green rather than grey leaves giving the whole mallee a green rather than blue-grey appearance. Although it is thought to be rare, this subspecies is likely to be more widespread in such a remote and inaccessible region, in similar sand plain habitats.

E. canescens subsp. *beadellii* on the Oak Valley to Tjuntjunjara track, Great Victoria Desert, S.A.

dried buds

Eucalyptus gypsophila Nicolle

Kopi Mallee

Genus equivalent: *Symphyomyrtus*; section: *Dumaria*; series: *Rufispermae*

NAME : From the Greek *gypsos* meaning 'gypsum', and *philos* meaning 'loving', referring to the common occurrence of this species on gypseous sand dunes around salt lakes.

DESCRIPTION : A mallee 3 to 6 metres tall. The bark is rough on the lower half of the stems, sub-fibrous and flaky then smooth, tan to grey and cream. The branchlets possess pith glands and are sometimes glaucous. The adult leaves are broad-lanceolate to lanceolate, dull and blue-green to blue-grey. The buds and fruits are sometimes glaucous, occurring in umbels of seven to eleven. The bud operculum is rounded to conical and smooth or lightly ribbed. The fruits are 6 to 10 mm long by 6 to 11 mm wide, smooth or very faintly ribbed. The flowers are white. Seeds are glossy, red-brown.

DISTRIBUTION & HABITAT : Widespread throughout the Great Victoria Desert, from west of the Gawler Ranges westwards, except the far northern part of the desert, extending into Western Australia as far as the Gascoyne region. It grows on sand dunes and sand plains overlying shallow, powdery limestone, often around or near small dry lakes (It is common around the Serpentine Lakes). Commonly associated species include *E. eucentrica, E. gongylocarpa, E. youngiana, E. concinna* and *E. eremicola.*

CULTIVATION & USES : Kopi mallee is rarely cultivated. It is not particularly ornamental but may be useful for shelter and erosion control in most soils, including those over limestone, in drier parts of the State. It will not tolerate waterlogging and is slow growing in Adelaide.

NOTES : Previously included in *E. striaticalyx*, a tree restricted to the northern goldfields of Western Australia. Distinguished from the closely related *E. canescens* (both subspp. *canescens* and *beadellii)* in the narrower adult leaves; smaller buds with a less ribbed operculum and the smaller, smoother fruits.

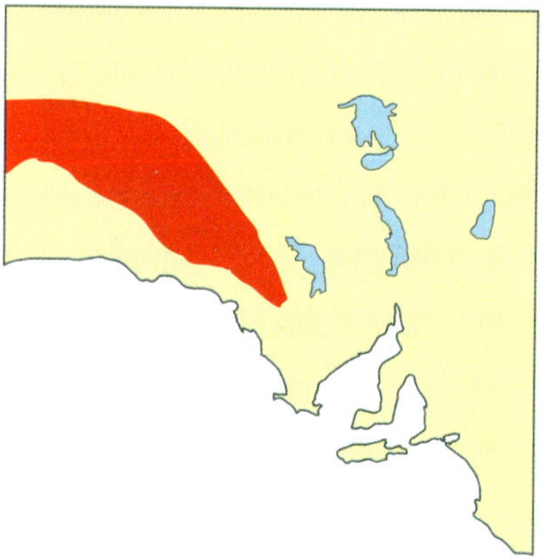

E. gypsophila on the Tarcoola to Ceduna track (Googes Track), Great Victoria Desert, S.A.

Eucalyptus flindersii Boomsma

Mallee Red Gum; South Australian Grey Mallee; Grey Mallee

Genus equivalent: *Symphyomyrtus*; section: *Exsertaria*; series: *Exsertae*

NAME : Commemorating the explorer Captain Matthew Flinders (1774 - 1814), after whom the Flinders Ranges were named.

DESCRIPTION : A mallee 1 to 5 metres tall. The bark is usually smooth and white to pinkish grey. A little rough bark is sometimes present on the lower stems of larger mallees. The juvenile leaves are ovate to rounded and green. The adult leaves are slightly glossy and green to slightly blue-green. Buds and fruits are in umbels of seven. The buds are variable in shape and size with a smooth, rounded to bluntly conical operculum (cap). Fruit size is also variable, becoming smaller to the north of its range. The flowers are white. The seeds are black (yellow in *E. camaldulensis*).

DISTRIBUTION & HABITAT : Restricted to the Flinders Ranges from the Yappala Range west of Hawker northwards to the Mawson Plateau (Freeling Heights) at the northern extremity of the Flinders Ranges, with outliers to the south on Devils Peak, the hills around Bibliando Station and Pualco Hill south of Yunta. It occurs on the upper slopes and summits of high mountains, usually on steep, windswept sites between rocks. It is usually the dominant species where it occurs. Associated species include *E. goniocalyx* subsp. *exposa*, *E. dumosa* and *E.* species 'Flinders Ranges'. Endemic to South Australia.

CULTIVATION & USES : In cultivation it forms a dense rounded shrub 2 x 2 metres in size, excellent for windbreaks and screens, however, it is rarely cultivated. It grows in most soils and tolerates some limestone. Plants are initially quick growing and have few pests and diseases. It may be useful as a windbreak or for erosion control, especially on rocky, steep sites.

NOTES : Similar to *E. gillenii*, differing in the rounded juvenile leaves and the flat-topped (rather than rounded), smaller fruits. The dominant mallee of the higher parts of the northern Flinders Ranges.

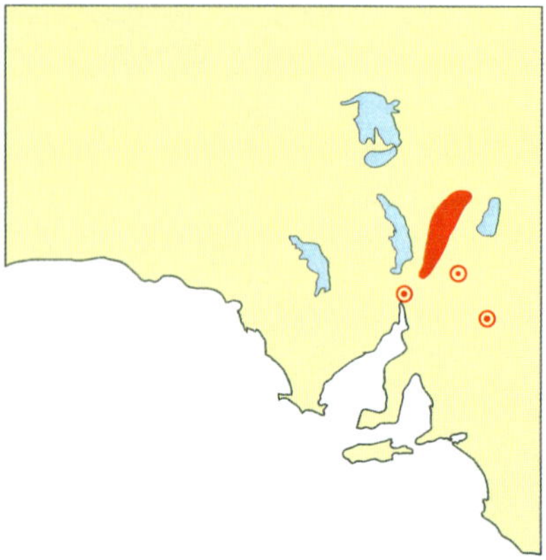

E. flindersii near Tanderra Saddle, Wilpena Pound Range, Flinders Ranges, S.A.

E. flindersii (in foreground) on Devils Peak looking north towards Duchmans Stern, Flinders Ranges, S.A.

Eucalyptus gillenii Ewart & Kerr

Mallee Red Gum; Mt Lindsay Mallee (SA); Mt Gillen Mallee (NT)

Genus equivalent: *Symphyomyrtus*; section: *Exsertaria*; series: *Exsertae*

NAME : Named after F.J. Gillen, the first postmaster at Alice Springs and after whom Mount Gillen in the MacDonnell Ranges, where the type of this species was collected, was named.

DESCRIPTION : A mallee 2 to 5 metres tall. The bark is smooth and grey to cream. The juvenile leaves are broad-lanceolate to lanceolate in shape. The adult leaves are dull to slightly glossy and green to slightly blue-green, with many small oil glands when held up into the light. The buds and fruits occur in umbels of seven. The flowers are white. The seeds are dark brown to black.

DISTRIBUTION & HABITAT : In South Australia known only from Mount Wooltarlinna and Mount Lindsay in the Birksgate Range in the far north-west of the State. It grows in dense thickets in crevices on massive rock domes where water accumulates, in places with stunted plants of *C. eremaea* subsp. *eremaea* and *E. intertexta*. *E. gillenii* does not grow on the sandy plains below these rock-faced mountains. It is more common in the MacDonnell Ranges of the Northern Territory and extends just into Western Australia on the Dean Range. It probably grows on other similar large rocky hills in the Birksgate Range, but several searches in the similar Everard Ranges have failed to find it there.

CULTIVATION & USES : *E. gillenii* is a rapid grower in cultivation, sometimes single-stemmed in sheltered situations, however, it usually forms a dense mallee suitable for windbreaks and wind erosion control. If prefers clay or loamy soils. It grows small enough to warrant use in home gardens.

NOTES : The South Australian populations of *E. gillenii* were once called *E. incurva*. *E. gillenii* is unlikely to be confused with any other eucalypt in the field. It is closely related to *E. flindersii*, differing in the longer, lanceolate juvenile leaves; the often larger, round-topped (rather than flat-topped) fruits and the northern distribution.

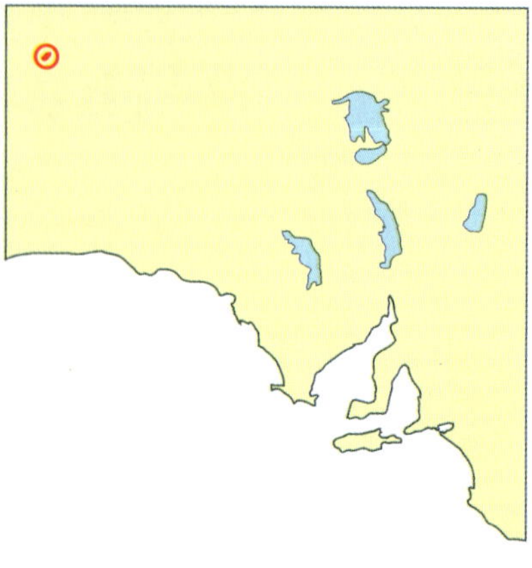

E. gillenii on Mount Wooltarlinna, Birksgate Range, Great Victoria Desert, S.A.

Eucalyptus camaldulensis Dehnh. var. *obtusa* Blakely

River Red Gum; Northern River Red Gum

Genus equivalent: *Symphyomyrtus*; section: *Exsertaria*; series: *Exsertae*

NAME : The varietal name is from the Latin *obtusus* meaning 'blunt or obtuse' referring to the rounded operculum compared to var. *camaldulensis*.

DESCRIPTION : A tree, often large and spreading, 8 to 25 metres tall. The bark is smooth (a little chunky rough bark may be present at base), powdery white or colourful with patches of grey, brown, cream and orange. The juvenile leaves are glaucous and broader than in var. *camaldulensis*. The adult leaves are dull to slightly glossy, light green to blue-green. The buds and fruits occur in umbels of seven to thirteen. The buds have a rounded to shortly conical operculum. The fruits are identical to var. *camaldulensis*. The flowers are white and seeds are pale yellow in colour.

DISTRIBUTION & HABITAT : Widespread in the north of the State, along major watercourses that are mostly dry. It occurs around the northern ranges where water accumulates and is common along the major watercourses draining into Lake Eyre (e.g. River Frome, Cooper Creek), Lake Torrens and Lake Frome. It is common in watercourses in the northern Flinders Ranges where it grades into var. *camaldulensis*. Also in N.S.W., Queensland, N.T and W.A.

CULTIVATION & USES : Commonly cultivated, especially overseas, it forms a large attractive tree. Some provenances are extremely fast growing, even in the south of the State, where it can be superior in growth rates to var. *camaldulensis*. It is suitable for farms and stations for woodlots and shade. It grows in most soils but it requires access to ground water or periodic flooding in dry parts of the State. It grows too large for general garden use.

NOTES : Differs from var. *camaldulensis* in the more colourful bark; the broader, glaucous juvenile leaves and especially the rounded operculum of the buds. The two varieties intergrade extensively and most populations in the northern Flinders Ranges are intermediate between the two varieties. An attractive, distinctive and very useful large tree in the north of the State.

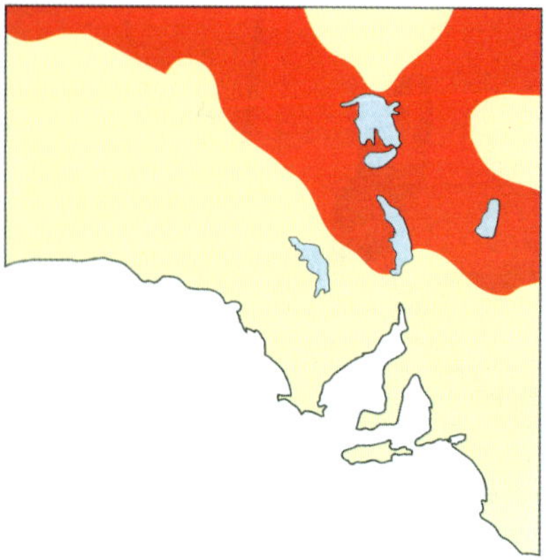

E. camaldulensis var. *obtusa* in Mundy Creek (in flood) near Lyndhurst, S.A.

Eucalyptus camaldulensis Dehnh. var. *camaldulensis*

River Red Gum; Murray Red Gum; Red Gum

Genus equivalent: *Symphyomyrtus*; section: *Exsertaria*; series: *Exsertae*

NAME : From Camaldoli, in Tuscany, Italy, where the tree from which the type material was taken was grown.

DESCRIPTION : A large tree with a spreading crown in open situations, 15 to 40 metres tall. The bark is mostly smooth, cream, grey and shades of tan, shedding in large plates. Some coarse rough bark may be present at the base of older trees. Adult leaves are thin and pendulous, dull to slightly glossy, green to blue-green. The small buds and fruits occur in umbels of seven to thirteen. Flowers are white, occurring in summer. Seeds are yellow, released soon after fruit maturity.

DISTRIBUTION & HABITAT : Widely distributed along the Murray River and along watercourses and floodplains in the South-east, the Flinders Ranges and the Mid-north, with disjunct occurrances on Yorke Peninsula, on western Eyre Peninsula and near Kingscote on Kangaroo Island. Also common along water-courses in Victoria, N.S.W. and southern Queensland. It always grows along the banks and floodplains of watercourses, usually in sandier soils than *E. largiflorens*. An exception is in the Mount Lofty Ranges where it also occurs on hillsides. The occurences on Eyre Peninsula and near Port Pirie occur on shallow soils over limestone.

CULTIVATION & USES : One of the most widely cultivated trees in Australia and overseas. Because it is naturally widespread, it adapts well to most areas, although it is best to choose local provenances. Its red timber is highly valued for fuelwood. It is suited for woodlots and shade but is too large for garden use. Because of its deep rooting system and open crown, other plants and pasture can be grown beneath this species.

NOTES : Probably the most well known eucalypt in Australia. Large, old trees are very characteristic and valued and many remnant trees are dotted throughout Adelaide. Similar to *E. camaldulensis* var. *obtusa*, differing in the longer, beaked operculum; the less colourful bark and the narrower, less glaucous juvenile leaves. The two varieties intergrade extensively in the northern Flinders Ranges.

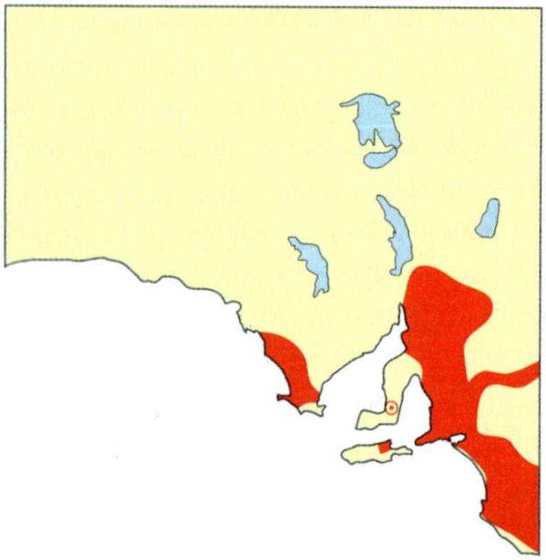

E. camaldulensis var. *camaldulensis* near Murray Town, southern Flinders Ranges, S.A.

Eucalyptus ovata Labill. var. *ovata*

Swamp Gum

Genus equivalent: *Symphyomyrtus*; section: *Maidenaria*; series: *Foveolatae*

NAME : From the Latin *ovatus* meaning 'ovate', referring to the shape of the adult leaves.

DESCRIPTION : A tree 6 to 20 metres tall usually with a single trunk. The bark is rough on the trunk and largest branches, coarsely fissured, grey to dark grey-brown, then smooth white to grey above. The adult leaves are broad (ovate to broad-lanceolate), thin and often wavy edged (undulate), glossy and dark green, with few or no oil glands when held up to light. The buds and fruits occur in umbels of seven. The buds are diamond-shaped, the operculum being pointed. The fruits are distinctly obconical (like an up-turned cone) in shape. The flowers are white.

DISTRIBUTION & HABITAT : Swamp Gum occurs in the lower South-east in broad, level, swampy areas, much of which has now been cleared. It also occurs in the southern Mount Lofty Ranges and on south-western Kangaroo Island (mainly within Flinders Chase National Park and Kelly Hill Conservation Park). It grows in broad seasonally swampy creeks, and low swampy valleys. In such waterlogged sites it may be the only tree present, although it may sometimes be associated with *E. viminalis*, *E. cosmophylla* and *E. paludicola*.

CULTIVATION & USES : Extremely fast growth rates can be achieved if water is assured and this eucalypt is very useful for seasonally waterlogged sites in sands or clays, where little else of substantial size will grow. It also grows well in well drained sites in high rainfall areas (> 700 mm annually). It forms an erect tree in cultivation, suitable for shade and fuelwood and, with good forms, timber production. It is a favourite food tree for koalas.

NOTES : Swamp gum is easily recognised by its broad but thin, glossy, green adult leaves and the small diamond-shaped buds in sevens and the obconical fruits, both much smaller than in *E. ovata* var. *grandiflora*. Both varieties of *E. ovata* are notable because of their swampy habitat.

E. ovata var. *ovata* in Tookayerta Creek, Fleurieu Peninsula, S.A.

Eucalyptus ovata Labill. var. *grandiflora*

Large-flowered Swamp Gum

Genus equivalent: *Symphyomyrtus*; section: *Maidenaria*; series: *Foveolatae*

NAME : The varietal name is from the Latin *grandis* meaning 'large' and *florus* meaning 'flowered' referring to the large flowers in comparison to var. *ovata*.

DESCRIPTION : A tree 6 to 18 metres tall. The bark is rough for 1 to 3 metres, flaky, grey, then smooth, grey and tan to cream. The adult leaves are often wavy (undulate), glossy and green, usually with many oil glands when held up to the light. The buds and fruits occur in umbels of seven. The buds are diamond-shaped with a pointed, beaked operculum. The fruits are obconical (inverted cone-shaped), and 8 to 10 mm long by 7 to 13 mm wide. The flowers are white.

DISTRIBUTION & HABITAT : Restricted to the south-east of the State in low, seasonally swampy areas in sandy to loamy soils. It grows from about Beachport to Penola and southwards, extending into adjacent Victoria as far east as Moonlight Head.

CULTIVATION & USES : Virtually unknown in cultivation. It would have the same general requirements for growth as *E. ovata* var. *ovata*.

NOTES : The glossy, broad, wavy leaves; the diamond shaped buds and the obconical fruits are distinctive for *E. ovata*. *E. ovata* var. *grandiflora* is very similar to *E. ovata* var. *ovata* but has much larger buds and fruits (fruits to 8 mm long and wide in var. *ovata*) and usually has many oil glands in the leaves. The two varieties grow in the same area but the populations are distinct and they are not known to intergrade. A coastal form of *E. ovata* occurring at Carpenter Rocks and Port MacDonnell is notable for its shorter, blunt rather than pointed operculum and may represent a further distinct taxon.

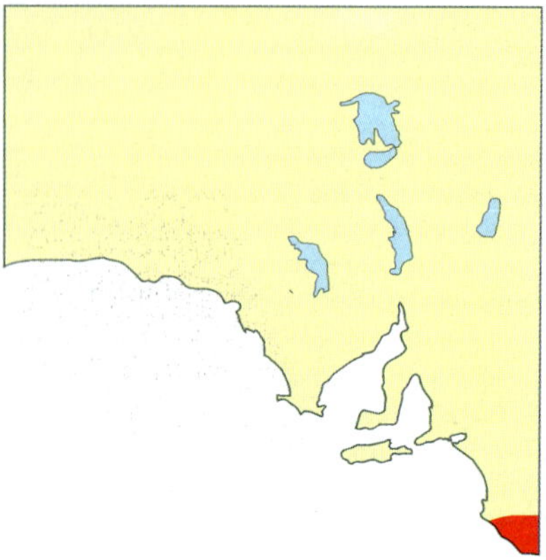

E. ovata var. *grandiflora* near Mount Richmond, Victoria.

Eucalyptus bicostata Maiden, Blakely & Simmonds

Southern Blue Gum; Eurabbie; Blue Gum; Victorian Blue Gum

Genus equivalent: *Symphyomyrtus*; section: *Maidenaria*; series: *Viminales*

NAME : From the Latin *bi* meaning 'two' and *costatus* meaning 'ribbed' referring to the two-ribbed buds and fruits.

DESCRIPTION : A gnarled, sprawling tree in South Australia, 5 to 16 metres tall, elsewhere a forest tree to 45 metres tall. The bark is ribbony rough at base for about a metre or two, then smooth, tan to grey and cream and ribbony. The juvenile leaves are conspicuous, being thin, grey and glaucous. Adult leaves are long, glossy and dark green. Buds and fruits occur in umbels of three and are sessile (lacking peduncles and pedicels), warty and glaucous. The fruits have two opposite ribs. The flowers are white.

DISTRIBUTION & HABITAT : Very rare in South Australia, known from a single population on hills near Mount Bryan, north of Burra. It occurs on the steep upper south facing slopes, above a zone of *E. leucoxylon* subsp. *pruinosa*, where snow and frosts in winter are relatively frequent. Common in the cool mountainous regions of New South Wales and Victoria.

CULTIVATION & USES : Commonly cultivated in Adelaide and the eastern states, it forms a large spreading tree with a compact, very dense crown, casting a heavy shade, beneath which little will grow. Suitable for shade and tall windbreaks, it prefers heavy soils in areas of high rainfall. Very fast growing in the ideal conditions. Young plants (the juvenile leaves) are susceptible to leaf eating caterpillars. The closely related *E. globulus* (from Victoria and Tasmania) is generally superior as a timber tree.

NOTES : Only recently discovered in South Australia, it is a distinctive tree, not likely to be confused with any other native to South Australia. The nearest native population to Mount Bryan is in the Otway Ranges of Victoria, over 600 km away. The buds and fruits in threes distinguish it from the commonly planted *E. globulus* (buds and fruits are singular), it also has broader, longer leaves and a more dense, rounded crown compared to *E. globulus*.

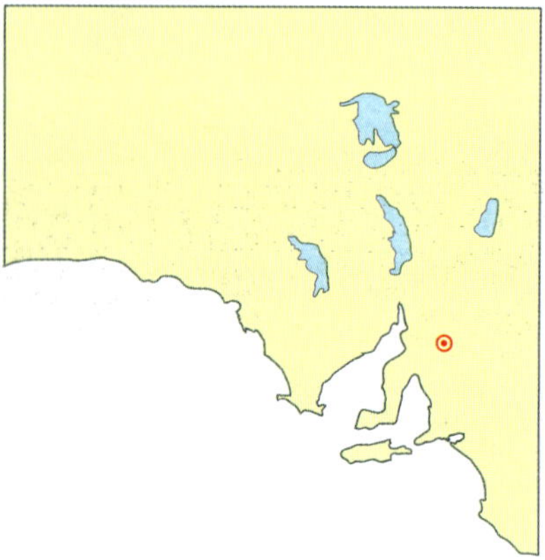

E. bicostata near Mount Bryan, east of Hallett, S.A.

juvenile leaves

Eucalyptus goniocalyx F. Muell ex Miq. subsp. *goniocalyx*

Long-leaved Box; Bundy; Olive-barked Box

Genus equivalent: *Symphyomyrtus*; section: *Maidenaria*; series: *Viminales*

NAME : From the Greek *gonia* meaning 'angle' and *calyx* (the calyx), referring to the slightly angled sides to the buds and fruits.

DESCRIPTION : A tree, sometimes of poor form and several stemmed, 6 to 20 metres tall. The bark is rough to the smallest branches, thick, fibrous and dark grey to grey-brown. The juvenile leaves are round, sessile, grey and glaucous, often conspicuous near mature trees on road verges. The adult leaves are glossy and dark green. The buds and fruits occur in groups of seven with individual buds and fruits being sessile (lacking pedicels). The lower half of the buds and fruits are somewhat angular. The flowers are white.

DISTRIBUTION & HABITAT : Of restricted and disjunct distribution in South Australia, occurring in the northern Mount Lofty Ranges in the hills north-east of Adelaide, such as along the River Torrens gorge and the hills around Clare to the Tothill Range, also in the southern Flinders Ranges near Wirrabara and from the Mount Remarkable area to Mountt Brown. It always occurs in hilly areas on the slopes and ridges, often on rocky areas in woodland, with species such as *E. leucoxylon* in the northern Mount Lofty Ranges and *E. cladocalyx* in the southern Flinders Ranges. It is more widespread in Victoria and New South Wales.

CULTIVATION & USES : The juvenile leaves of this species are often picked from the wild and used in floral arrangements in Adelaide. It is rarely cultivated, but suitable for any well drained soil in areas of moderate to high rainfall. Useful for exposed, rocky sites. Plants have a moderate growth rate and are suitable for shelter, shade and woodlots.

NOTES : The rough, fibrous bark; glossy, dark green leaves and small, sessile buds and fruits distinguish this subspecies from other South Australian eucalypts. Closely related to *E. goniocalyx* subsp. *exposa,* differing in the non-glaucous adult parts and the larger, glossy, dark green adult leaves.

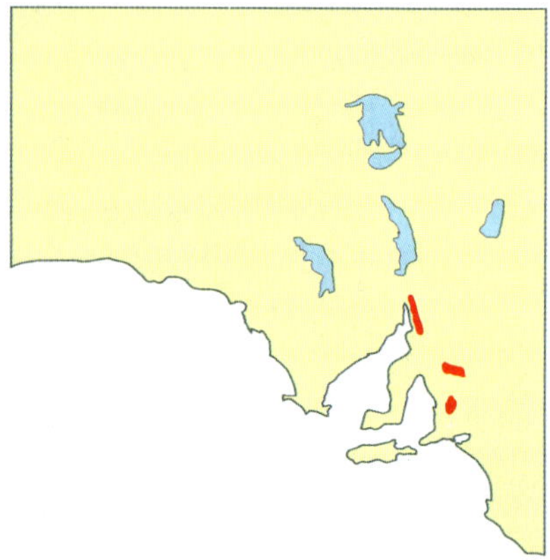

E. goniocalyx subsp. *goniocalyx* in Para Wirra Recreation Park, Mount Lofty Ranges, S.A.

juvenile regrowth

Eucalyptus goniocalyx F. Muell ex Miq. *subsp. exposa* Nicolle ined.

South Australian Mountain Box

Genus equivalent: *Symphyomyrtus*; section: *Maidenaria*; series: *Viminales*

NAMES : The undescribed subspecific name is from the English *exposed*, refering to the very exposed sites on which it grows.

DESCRIPTION : Usually a mallee, sometimes developing into a stunted tree to 7 metres tall. The bark is variable, smooth grey to cream in smaller mallees or with some rough, thin, finely fissured grey bark on the lower stems on larger plants. The twigs, buds and fruits are slightly to heavily glaucous. The juvenile leaves are sessile, round, opposite and grey-glaucous. Adult leaves are dull to slightly glossy, light green to bluish. The buds and fruits occur in umbels of seven. The flowers are white.

DISTRIBUTION & HABITAT : Restricted to high mountain peaks of the Elder and Wilpena Pound Ranges in the northern Flinders Ranges of South Australia. It grows on exposed, rocky sites, only on peaks and ridges higher than 900 metres, for example St Mary Peak, Pompey Pillar, Point Bonney and Mount Aleck. It is the dominant species on these peaks or sometimes grows associated with *E. flindersii*. Endemic to South Australia.

CULTIVATION & USES : Virtually unknown in cultivation, trials are being undertaken to determine if the small mallee habit is retained in cultivation. It requires a well drained soil and may be useful for shelter on exposed, rocky sites.

NOTES : A recently recognised taxon closely related to both *E. nortonii* and *E. goniocalyx*, but probably evolved from the latter. From *E. goniocalyx* subsp. *goniocalyx* it differs in the smaller habit and often smooth bark, especially on the branches; the smaller, more crowded juvenile leaves; the duller, much smaller adult leaves and the glaucous twigs, buds and fruits. Differs from *E. nortonii*, which is native to New South Wales and Victoria, in its small mallee habit and often smooth bark; the smaller and more crowded juvenile leaves; the smaller, greener adult leaves; the weaker glaucousness and usually smaller buds and fruits.

E. goniocalyx subsp. *exposa* on Mount Aleck looking north along the Elder Range with the Wilpena Pound Range in the distance, northern Flinders Ranges, S.A.

E. goniocalyx subsp. *exposa* on Point Bonney, Wilpena Pound Range, northern Flinders Ranges, S.A.

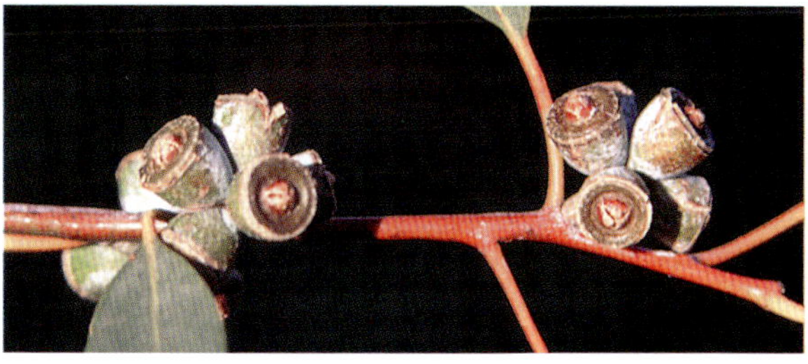

Eucalyptus viminalis Labill. subsp. *cygnetensis* Boomsma

Rough-barked Manna Gum

Genus equivalent: *Symphyomyrtus*; section: *Maidenaria*; series: *Viminales*

NAMES : This subspecies was named after the Cygnet River on Kangaroo Island, where the type specimen was collected.

DESCRIPTION : A tree, sometimes of poor, spreading form, 6 to 20 metres tall. Trees are generally taller and straighter in the lower South-east and adjacent parts of Victoria. The bark is rough on at least the lower half of the tree, grey-brown, then smooth on upper part or only on the small branches. The juvenile leaves are glossy and green. Adult leaves are usually glossy and green. In the Coorong and lower Murray Mallee areas, the adult leaves are generally less glossy, broader and thicker than elsewhere. Buds and fruits occur in groups of mostly sevens (some threes may be present). The flowers are white.

DISTRIBUTION & HABITAT : Scattered in South Australia in the Mount Lofty Ranges and on Kangaroo Island, extending east to very scattered populations in the upper South-east and lower Murray Mallee areas. Also occurs south-east to Mount Gambier, where it becomes more common, and east to Port Phillip Bay in Victoria. An isolated population occurs on limestone near Port Lincoln on Eyre Peninsula. It usually occurs on sandy soils on moist sites, e.g. creeks and sandy depressions, but occurs on hills in heavier soils in parts of the Mount Lofty Ranges.

CULTIVATION & USES : Occasionally cultivated, it is generally inferior to subsp. *viminalis* in growth rate and form, but is more suited to drier, infertile, sandy sites. It is a fast growing tree that is attractive in cultivation, being suitable for parks, roadside plantings and farms, for shade, tall shelter and water erosion control. A favourite food tree for koalas.

NOTES : Distinguished from *E. viminalis* subsp. *viminalis* by the poorer form; rough bark on the trunk and larger branches and the buds and fruits in groups of mostly sevens. The two subspecies intergrade in the central Mount Lofty Ranges, with trees having a short stocking of rough bark for up to 3 metres.

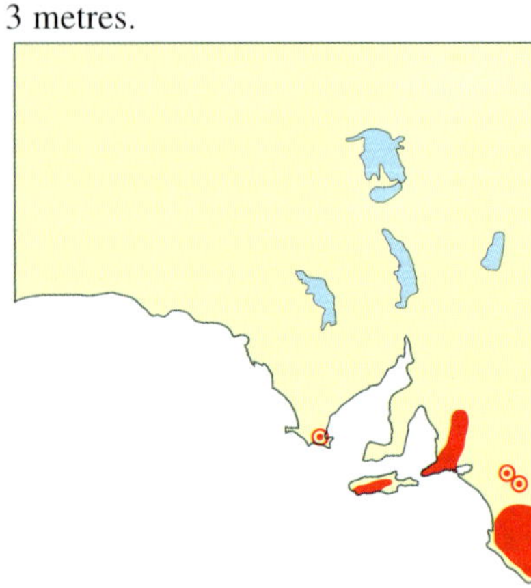

E. viminalis subsp. *cygnetensis* near Currency Creek, Fleurieu Peninsula, S.A.

Eucalyptus viminalis Labill. subsp. *viminalis*

Manna Gum; Ribbon Gum; White Gum

Genus equivalent: *Symphyomyrtus*; section: *Maidenaria*; series: *Viminales*

NAME : From the Latin *viminalis* meaning 'viminous' or 'osier-bearing', referring to the adult leaves resemblance to those of the osier willow.

DESCRIPTION : A tree 15 to 40 metres tall in South Australia. The bark is often rough, fibrous and brown for a metre or two, otherwise smooth and white to grey. The juvenile leaves are opposite, lanceolate in shape, glossy and green. Adult leaves are also glossy and green. The buds and fruits occur in umbels of three. The flowers are white.

DISTRIBUTION & HABITAT : Rare in South Australia, restricted to the higher parts of the central Mount Lofty Ranges where it grows on moist slopes, sometimes with *E. obliqua* or *E. dalrympleana*. It is a common tree of the wet forests of Victoria (from the Otway Range eastwards), New South Wales and Tasmania.

CULTIVATION & USES : Commonly cultivated, this tree is extremely fast growing in areas of high rainfall. It is recommended for farms and other broadscale planting for shade, woodlots and ornamental purposes. In excellent conditions it is cultivated for pulp and timber production. It requires assured rainfall in any well-drained soil. The straight form and crown of narrow, green, pendulous leaves is ornamental, but the tree is unsuitable for gardens due to its large size. It is a favourite food tree of koalas.

NOTES : One of Australia's tallest trees, attaining heights of up to 90 metres in Tasmania. It is often confused with *E. dalrympleana*, however, its narrow, green juvenile leaves (compared to round, grey-green juvenile leaves in *E. dalrympleana*) distinguish the two. *E. viminalis* subsp. *viminalis* also has less colourful smooth bark; green rather than pink adult new growth and less wavy adult leaves than *E. dalrympleana*. *E. viminalis* subsp. *viminalis* can be distinguished from subsp. *cygnetensis* by its taller form, mostly smooth bark and consistently three-flowered buds and fruits, however the two subspecies intergrade in drier parts of the central Mount Lofty Ranges.

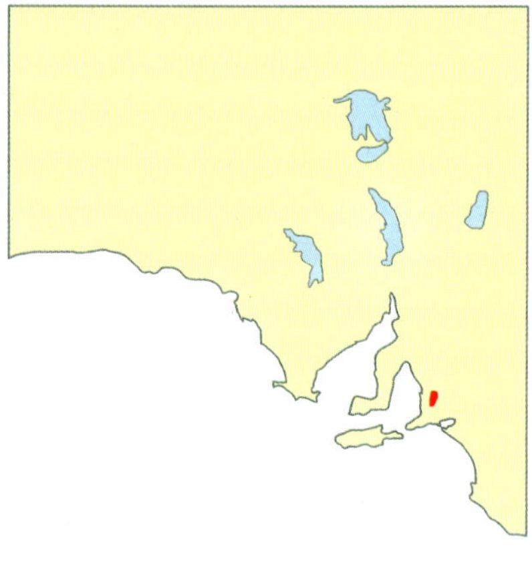

E. viminalis subsp. *viminalis* on Mount Osmond, Mount Lofty Ranges, S.A.

juvenile leaves

Eucalyptus dalrympleana Maiden subsp. *dalrympleana*

Mountain Gum; White Gum

Genus equivalent: *Symphyomyrtus*; section: *Maidenaria*; series: *Viminales*

NAME : Named in honour of Richard Dalrymple Hay (1861 - 1943), former Chief Commissioner of Forests in New South Wales.

DESCRIPTION : Usually a straight-trunked tree 15 to 45 metres tall, but smaller (to about 10 metres) near Mt. Compass and Parawa. In the eastern states it grows up to 60 metres tall. The bark is completely smooth, white and grey to dark pink. The juvenile leaves (often present underneath mature trees) are rounded and dull, grey-green. The adult leaves are slightly glossy, green and usually wavy (undulate). The buds and fruits are in umbels of three. The inconspicuous flowers are white.

DISTRIBUTION & HABITAT : Occurs in the higher parts of the Mt Lofty Ranges, from near Gumeracha to about Mylor, with a few small isolated occurances to the south in more swampy areas in and around Kyeema Conservation Park, near Mt Compass and around Parawa. It is common along the freeway between Stirling and Bridgewater. It occurs in tall forest sometimes with *E. obliqua* and *E. viminalis* subsp. *viminalis*. It is more common in New South Wales, Victoria and Tasmania.

CULTIVATION & USES : A very fast growing tree of good form, it prefers heavy but well drained soils in cool areas and it demands high rainfall. It tolerates frosts and snow. It is recommended for woodlots or timber plantations in the Mount Lofty Ranges. It grows far too large for home garden use.

NOTES : Once confused with *E. rubida* (candlebark), *E. dalrympleana* differs in the non-glaucous juvenile leaves, twigs, leaves, buds and fruits and the taller habit. *E. rubida* only occurs in Victoria, New South Wales and Tasmania. Mountain Gum is a stately tall tree, now quite rare in South Australia as it is mainly restricted to cold sites on fertile, highly productive soils. It is similar to *E. viminalis* subsp. *viminalis* differing in the rounded, dull, grey-green juvenile leaves; the more colourful smooth bark and, in South Australia, the usually taller habit.

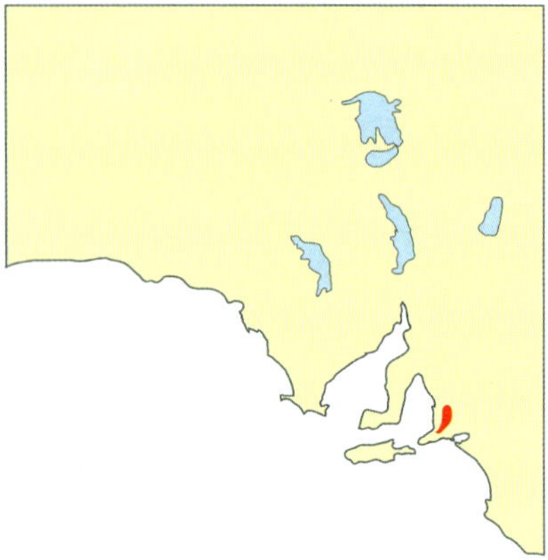

E. dalrympleana subsp. *dalrympleana* on Mount George Road, Bridgewater, Mount Lofty Ranges, S.A.

juvenile leaves

Eucalyptus species 'Carpenter Rocks'

Mallee Manna Gum

Genus equivalent: *Symphyomyrtus*; section: *Maidenaria*; series: *Viminales*

NAME : An unnamed species related to *E. splendens,* known only from Carpenter Rocks near Millicent.

DESCRIPTION : A straggly, low tree, often several stemmed and mallee-like, 2 to 5 metres tall. The bark is rough to the small branches, thin, flaky to slightly tessellated, light grey. The smooth bark above is grey to cream. Juvenile leaves are glossy, green and sometimes crenulate. Adult leaves are glossy and light green to green. The buds and fruits are sessile (lacking individual pedicels) and occur in umbels of seven. The buds have a rounded operculum. (compare to *E. viminalis* subsp. *cygnetensis*). The flowers are white.

DISTRIBUTION & HABITAT : A very rare eucalypt, known only from near Carpenter Rocks in the State's South-east. Only a few small populations about 1 km from the coastline in a conserved patch of scrub are known. It may have been more common before massive vegetation clearance in the area. It is possible that more populations may be found in nearby similar areas. It grows on low limestone in orange clay-loam with much outcropping limestone in dense coastal scrub with *E. obliqua* and *E. ovata* nearby. Endemic to South Australia.

CULTIVATION & USES : Rare in cultivation, but when cultivated it forms an attractive rounded tree about 10 metres tall. It may be very useful for difficult coastal sites on shallow soils over limestone in areas of rainfall above about 600 mm annually.

NOTES : A recently recognised species that is very rare in the wild. It is superficially similar to, and can be easily confused with *E. viminalis* subsp. *cygnetensis*, but differs mainly in the wider, rounded to elliptical, slightly crenulate juvenile leaves; obtuse operculum and the lower, straggly habit. It always grows on low limestone rises on the coast, while *E. viminalis* subsp. *cygnetensis* prefers deeper, sandy soils and rarely grows on limestone outcrops. *E.* species 'Carpenter Rocks' is probably closely related to *E. splendens*, from Victoria, which differs in its taller habit; smaller buds and fruits with longer peduncles and pedicels and a pointed operculum.

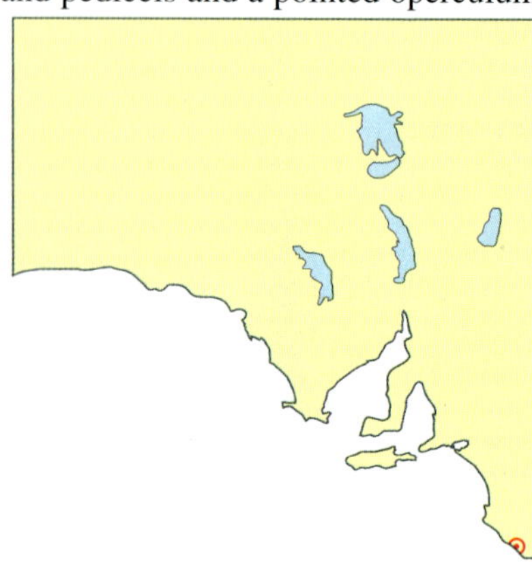

E. species 'Carpenter Rocks' near Carpenter Rocks, S.A.

Eucalyptus coolabah Blakely & Jacobs subsp. *arida* (Blakely) L. Johnson & K. Hill

Coolabah; Coolibah

Genus equivalent: *Symphyomyrtus*; section: *Adnataria*; series: *Striolatae*

NAME : Coolabah is the Aboriginal name for this species and the subspecific name is from the Latin *aridus* meaning 'dry' referring to its arid habitat.

DESCRIPTION : A spreading tree, sometimes several-stemmed, 4 to 12 metres tall. The bark is rough on the trunk and lower branches, finely fissured, grey to brown, then smooth and white above. Twigs, buds and fruits are non-glaucous. The adult leaves are dull, blue-green. The very small buds and fruits occur in umbels of seven to eleven and form larger terminal clusters. The fruits have exserted valves when dry. The flowers are white.

DISTRIBUTION & HABITAT : Widely distributed and of scattered occurrence in the north-east of the State, occurring mostly within the Lake Eyre Basin. It extends south to around Lake Frome and on the west side of Lake Torrens to west of Mabel Creek. It grows along creeks and on flood-plains and sometimes between widely spaced sand dunes (in the Strzelecki Desert). It is the dominant tree of the channel country including the Cooper Creek. It is occasionally associated with *E. camaldulensis* var. *obtusa*. *E. coolabah* subsp. *arida* also grows in adjacent parts of Northern Territory, Queensland and New South Wales.

CULTIVATION & USES : Occasionally cultivated in drier parts of the State where it forms an attractive shady tree suitable for parks and farms. It also grows quite well in Adelaide, where it is slower growing.

NOTES : The 'Dig Tree'of Burke and Wills, occurring in far south-western Queensland, is the most famous tree of this species. This species is easily distinguished from other South Australian species by its smooth white upper bark above rough bark; the dull, blue-green leaves and the tiny buds, flowers and fruits. *E. coolabah* subsp. *arida* is very similar to subsp. *coolabah*, differing in its thicker adult leaves, more rounded operculum and its non-glaucous twigs, leaves, buds and fruits. A single record of *E. coolabah* subsp. *coolabah* is known at Lake Perigunda, Cooper Creek, but it is otherwise restricted to New South Wales and Queensland.

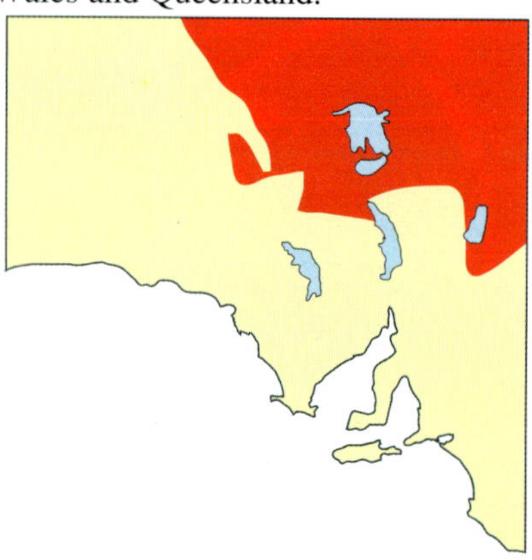

E. coolabah subsp. *arida* on the Stuart Highway at the Palmer River crossing, N.T.

Eucalyptus lansdowneana F. Muell. & J.E. Brown

Red-flowered Mallee Box; Crimson mallee

Genus equivalent: *Symphyomyrtus*; section: *Adnataria*; series: unnamed

NAME : Named in honour of Thomas Lansdowne Brown (1860 - 1931), one of the first collectors of the species.

DESCRIPTION : A thin-stemmed, scraggy mallee, 2 to 5 metres tall, with a very open crown that is finally pendulous. The bark is smooth and white to grey, although a little rough fibrous bark may be present for a metre or so at the base of large mallees. The adult leaves are long and often curved (falcate), very glossy and green to yellow-green. The buds and fruits occur in umbels of seven in terminal clusters or along leafless sections of the twigs. The buds and fruits are noticeably angled or with weak ribbing. The flowers are bright red fading to dark pink with age (very rarely white flowers are recorded).

DISTRIBUTION & HABITAT : Restricted to the higher south-western part of the Gawler Range between Minnipa and Yardea, in an area bounded by Kododo Hill, Conical Hill, Paney Bluff and the hills south-west of Scrubby Peak. It occurs on rocky hillsides and ridges in low shrubland, with species such as *E. porosa* and *E. phenax*. Endemic to South Australia.

CULTIVATION & USES : Sometimes cultivated as an unusual and ornamental plant, with its drooping habit and masses of red flowers. It can become straggly with age and cutting back to ground level every 8 years or so produces a more attractive plant with better flower production. A recommended eucalypt for small gardens in warm to hot areas of the State in most well-drained soils. Apart from ornamental uses it has limited value because of its small, poor form.

NOTES : A very distinctive mallee because of its unusual, drooping habit; the large, glossy, yellow-green leaves and the red flowers. It differs from *E. albopurpurea*, which has white, pink or purple flowers, in the thin-stemmed habit; the longer, narrower, yellow-green leaves; the red flowers and the larger buds and fruits.

E. lansdowneana on Scrubby Peak, Gawler Ranges, S.A.

Eucalyptus intertexta R.T. Baker

Gum-barked Coolibah; Smooth-barked Coolibah

Genus equivalent: *Symphyomyrtus*; section: *Adnataria*; series: *Buxeales*

NAME : From the Latin *inter* meaning 'between' and *textus* meaning 'tissue', referring to the grain of the timber of this species which is reportedly difficult to split.

DESCRIPTION : Usually a tree up to 18 metres tall, often several-stemmed, reduced to a mallee 2 to 4 metres tall on the rocky slopes of the Everard Ranges. The bark is rough on the lower half of the stems, finely fibrous and sometimes wavy, grey, then smooth, white to pale grey above. The adult leaves are dull, green to grey-green. The small buds and fruits occur in umbels of seven that group together into larger groups of buds towards the ends of the branchlets. The fruits are barrel-shaped. The flowers are white. Plants in the Everard Ranges and rarely in the Flinders Ranges have glaucous twigs, buds and fruits.

DISTRIBUTION & HABITAT : Two large, disjunct occurrences are known in South Australia; in the northern Flinders Ranges extending to the Olary Ridges and in the northern ranges such as the Birksgate, Musgrave and Everard Ranges. There is also a third smaller occurrence in the hills and creeks on north-eastern Eyre Peninsula west of Port Augusta. In all three locations it grows in mainly dry water-courses and on the slopes of rocky hills. It also grows in New South Wales and Queensland, where it commonly grows into a larger tree. In the north-west of South Australia it extends into adjacent areas of Western Australia and Northern Territory.

CULTIVATION & USES : Only occasionally cultivated. It would be very useful for a shade tree in drier parts of the State. The timber is durable and has good burning qualities and it is suitable for woodlots. Plants have a slow to moderate growth rate. It will grow in most soils including those with some limestone.

NOTES : Distinguished by its combination of the lower rough bark then smooth branches; the dull, usually blue-green adult leaves; the small buds and the small, barrel-shaped fruits.

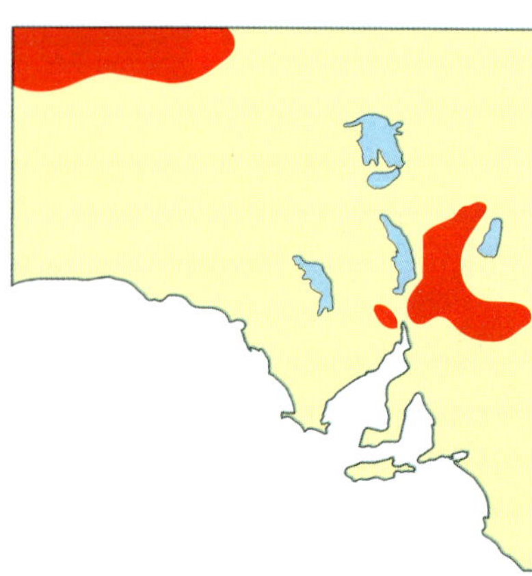

E. intertexta near the junction of Eyre and Lincoln Highways, northern Eyre Peninsula, S.A.

Eucalyptus sparsa Boomsma

Northern Ranges Box

Genus equivalent: *Symphyomyrtus*; section: *Adnataria*; series: *Buxeales*

NAME : From the Latin *sparsus* meaning 'sparse', referring to its very scattered occurrence.

DESCRIPTION : A spreading mallee 2 to 6 metres tall, often occurring in small clumps. It rarely develops into a low, spreading tree to 6 metres tall. The bark is rough on the lower part or over most of the stems, coarse, loose, grey-brown to almost black then smooth and cream above. Small mallees may be mostly smooth-barked. The adult leaves are very glossy and green. The small buds and fruits occur in umbels of seven in terminal clusters. The flowers are white, usually occurring in summer.

DISTRIBUTION & HABITAT : Rare and scattered in the far north-western part of the State and adjacent parts of Western Australia and Northern Territory. It favours red sands immediately below large hills, where rain runoff contributes to extra soil moisture. It also grows on sand plains, but always within sight of rocky hills or ranges. It occurs on and around the Everard, Mann, Birksgate and Musgrave Ranges. It always occurs in small clumps, often isolated from one another by many kilometres. Associated species include *C. eremaea* subsp. *eremaea, E. oxymitra* and *E. intertexta*.

CULTIVATION & USES : Suitable for a hot and sunny position in sandy to clay soils in areas of low rainfall. It has been tried in Adelaide, but is unreliable. It may be more successful in northern parts of the State. *E. sparsa* reshoots vigorously from ground level after fire or damage, producing a bushier mallee.

NOTES : The very glossy broad leaves make this species noticeable and easily distinguishable from *E. intertexta*, the only other box (section *Adnataria*) species that may grow with *E. sparsa*. The small, barrel-shaped, thin rimmed fruits identifies it from other glossy-leaved desert species such as *E. mannensis* and *E. concinna*.

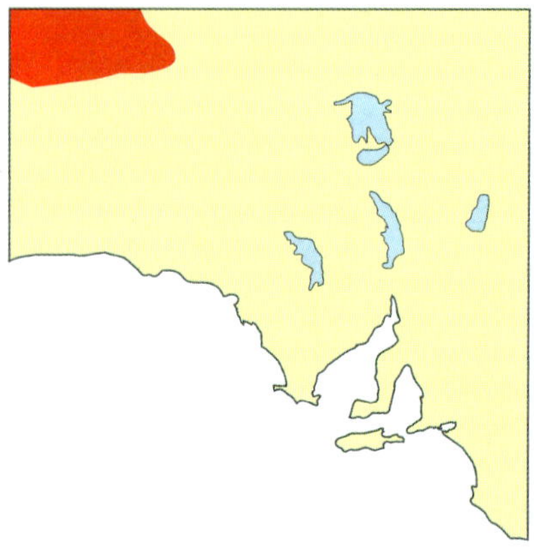

E. sparsa at Ronald Well, Everard Ranges, S.A.

E. sparsa near Moolalpinna Hill, northern Great Victoria Desert, S.A.

Eucalyptus behriana F. Muell.

Bull Mallee; Broad-leaved Box; Broad-leaved Mallee Box

Genus equivalent: *Symphyomyrtus*; section: *Adnataria*; series: *Buxeales*

NAME : Named after H.H. Behr (1818 - 1904), a botanical collector in South Australia.

DESCRIPTION : Varies from a tall mallee to 10 metres tall on Eyre Peninsula and in parts of New South Wales, to a small, often several-stemmed tree to 12 metres elsewhere. The bark is rough on the lower or most of the stems, moderately fissured, grey to dark brown, then smooth above, grey to olive green. The adult leaves are glossy, blue-green at first, soon maturing dark olive-green. The intramarginal vein is well in from the leaf edge. The buds and fruits occur in umbels of three to seven in large terminal groups on the ends of the branches. The flowers are white, occurring mainly in summer.

DISTRIBUTION & HABITAT : Known from three disjunct areas in South Australia. On the lower Eyre Peninsula in the Cummins area where it occurs in pure stands in lower areas than the surrounding mallee scrub of species such as *E. peninsularis, E. calycogona* and *E. odorata*. In the northern Mount Lofty Ranges from about Roseworthy to Riverton it is scattered in undulating areas in low open woodland of *E. leucoxylon* and *E. odorata*. It is more common in central western Victoria and extends just into South Australia north of Bordertown where it occurs in pure stands in low areas surrounded by mallee species and *E. leucoxylon*. subsp. *pruinosa*. It also occurs in the West Wyalong area of New South Wales.

CULTIVATION & USES : Little known in cultivation. Trials show it to have a moderate growth rate and although it prefers heavy soils, it grows well in sandy soils. It is somewhat ornamental and is suitable for wind-breaks and erosion control.

NOTES : The small, sessile buds and fruits in large terminal groups and the thick, ovate, glossy, dark green leaves distinguish this species from other box (section *Adnataria*) eucalypts. It usually occupies low flat areas similar to the habitat of *E. porosa*, with which it can be confused. *E. porosa* differs, among other things, in the non-sessile and larger buds and fruits and the thinner, often narrower adult leaves.

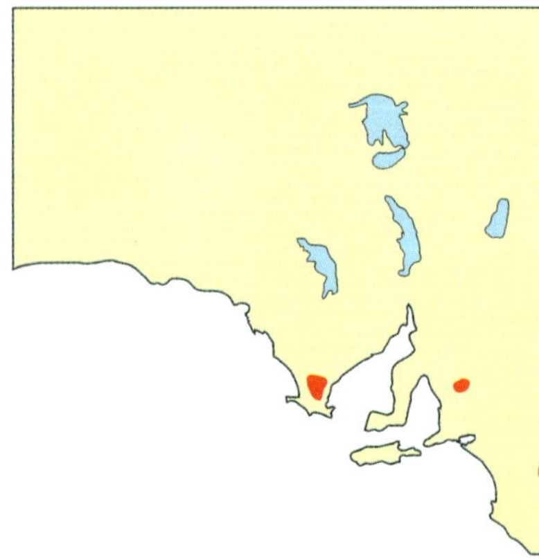

E. behriana on the Cummins to Yallunda Flat road, Eyre Peninsula, S.A.

Eucalyptus largiflorens F. Muell.

Black Box; River Box

Genus equivalent: *Symphyomyrtus*; section: *Adnataria*; series: *Buxeales*

NAME : From the Latin *largus* meaning 'abundant' and *florens* meaning blooming', referring to the plentiful flowers. The name is somewhat misleading as the flowers are quite small and usually inconspicuous.

DESCRIPTION : A tree 6 to 15 metres tall, with an open to moderately dense crown. The bark is completely rough, finely fissured and dark grey to almost black. The juvenile leaves are pendulous, long and narrow, often curved and blue-grey. Adult leaves are always dull and usually grey-green or rarely green. The buds and fruits occur in umbels of seven to eleven which form larger terminal groups. The flowers are white, but sometimes age to pink.

DISTRIBUTION & HABITAT : Black box is common on heavy clay soils along the floodplains of the Murray River as far downstream as Clayton and north in very scattered populations as far as the southern Strzelecki Desert, extending east into Victoria, New South Wales and southern Queensland. On lighter soils closer to water courses it is replaced by *E. camaldulensis* var. *camaldulensis*. It also occurs on the floodplains of the River Gawler in the Two Wells area and in black clay depressions around Bordertown.

CULTIVATION & USES : Often cultivated, it grows rapidly in heavy soils given ample water, especially in warmer parts of the State. It forms an attractive spreading tree with attractive, dark, rough bark and bluish foliage. It is suitable for shade, woodlots and water erosion control and is useful for areas prone to periodic flooding.

NOTES : A distinctive box, distinguished by its combination of floodplain habitat; dark, rough bark and usually grey-green adult leaves. The juvenile leaves which are often present near the mature trees will also distinguish this species from other box (section *Adnataria*) species. In the Milang and Angle Vale areas it may grow with the boxes *E. odorata* and *E. porosa*, which both have shorter, green juvenile leaves; glossy, green adult leaves and larger fruit than *E. largiflorens*.

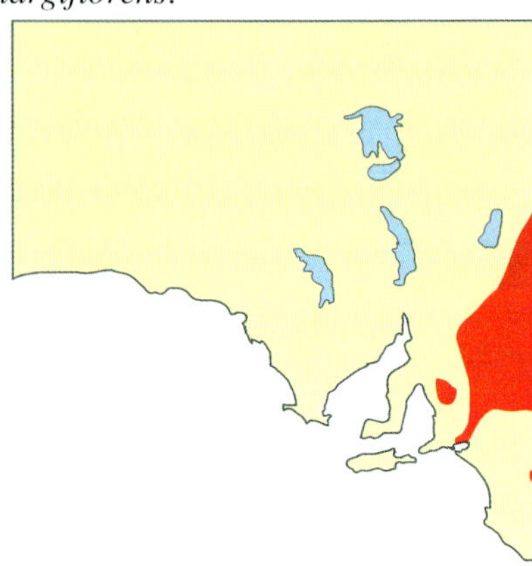

E. largiflorens near the Murray River at Loxton, S.A.

E. largiflorens near Wentworth, N.S.W.

Eucalyptus albens Benth.

White Box

Genus equivalent: *Symphyomyrtus*; section: *Adnataria*; series: *Moluccanae*

NAME : From the Latin *albens* meaning 'whitened' referring either to the whitish bark, or the glaucous white branchlets.

DESCRIPTION : A tree 10 to 18 metres tall. The bark is rough to the smaller branches, finely fissured, light grey, sometimes with whitish, bleached patches. The juvenile leaves are large, rounded to ovate, dull grey and sometimes glaucous. The adult leaves are dull, blue-green to greyish and sometimes glaucous. The branchlets, buds and fruits are glaucous. Buds and fruits are in umbels of seven to eleven in terminal clusters. The flowers are white.

DISTRIBUTION & HABITAT : Rare and of restricted distribution in South Australia, known only in the southern Flinders Ranges between Melrose and the Port Germein to Murray Town road, being common and best seen along Survey Road south of Melrose. Here it grows in hilly country on the hillsides as well as in creeklines, often with *E. microcarpa* and *E. camaldulensis* var. *camaldulensis*. It is more common and widely distributed in Victoria, New South Wales and far south-eastern Queensland.

CULTIVATION & USES : Occasionally cultivated, plants have a moderate growth rate and are attractive, particularly while young. White box is a useful shade tree and a very valuable honey tree and is also useful for woodlots, the timber, like all the boxes, having excellent burning qualities.

NOTES : Similar to the closely related *E. microcarpa*, with which it intergrades in the southern Flinders Ranges, differing in the greyer juvenile and adult leaves; the glaucous branchlets, buds, fruits and often leaves and the somewhat larger buds and fruits.

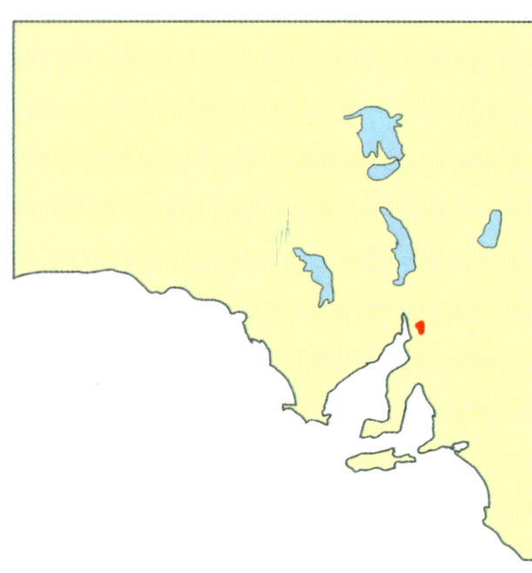

E. albens south of Melrose, southern Flinders Ranges, S.A.

Eucalyptus microcarpa (Maiden) Maiden

Grey Box; Green box

Genus equivalent: *Symphyomyrtus*; section: *Adnataria*; series: *Moluccanae*

NAME : From the Greek *micros* meaning 'small' and *carpos* meaning 'fruit', referring to the relatively small fruits of this species.

DESCRIPTION : A tree 6 to 20 metres tall, although reduced to a large, erect-stemmed mallee on some steep rocky sites near Adelaide. The bark is rough to the small branches, finely fissured and tightly held, grey. The juvenile leaves are broad (usually ovate) and grey-green. Adult leaves are glossy, green to olive green. The buds and fruits occur in umbels of seven, forming larger terminal groups on the ends of the twigs. The fruits are barrel-shaped. The flowers are white.

DISTRIBUTION & HABITAT : Restricted to three disjunct populations in South Australia. In the southern Flinders Ranges, especially around Melrose but extending north to the Quorn area; on the southern Adelaide plains and western slopes of the Adelaide hills south to Willunga and in the Bordertown to Naracoorte area in the south-east of the State where it continues into Victoria, New South Wales and southern Queensland. It prefers heavy soils such as clays and is sometimes associated with *E. camaldulensis* var. *camaldulensis* on flat sites.

CULTIVATION & USES : Occasionally cultivated, it prefers heavy clay soils but is adaptable. The timber has excellent burning qualities and it is worth planting for woodlots and shade although plants are generally not quick growing. Also a good honey tree. Old, remnant trees of this species are scattered in the southern and eastern suburbs of Adelaide.

NOTES : The three main populations in South Australia, as stated above, differ from one another slightly in juvenile and adult characteristics but here they are all included in *E. microcarpa*. As a whole they are often confused with *E. odorata* but *E. microcarpa* differs in the broader, greyish juvenile leaves; the usually taller habit; the more compact, finely fissured bark; the terminal flower clusters and the barrel-shaped fruits.

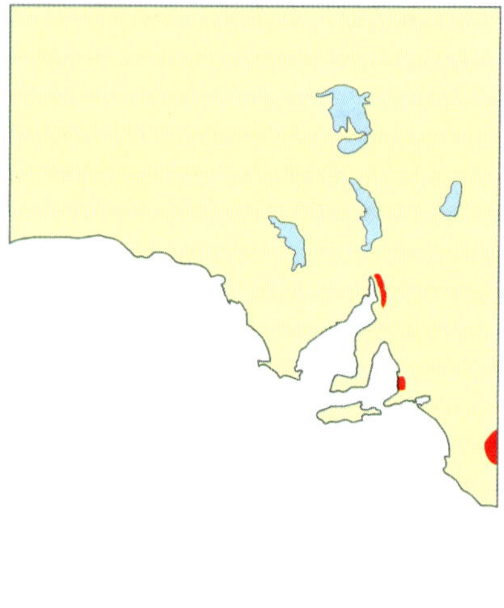

E. microcarpa in Pimpala Conservation Reserve, Morphett Vale, S.A.

Eucalyptus odorata Behr

Peppermint Box

Genus equivalent: *Symphyomyrtus*; section: *Adnataria*; series: *Subbuxeales*

NAME : From the Latin *odoratus* meaning 'having a smell', referring to the scent of the leaves when crushed (common to many eucalypts).

DESCRIPTION : Varies from a mallee to a spreading tree, 2 to 12 metres tall. The bark is rough on the trunk(s) and usually branches, thick, fibrous and grey-brown, then smooth on the smallest branches. Small 'whipstick' mallees occurring as an understorey to larger eucalypts may be completely smooth barked. The adult leaves are dull, blue-green when new, soon maturing glossy and green with age. The buds and fruits occur in umbels of seven to eleven. The fruits are cup-shaped to slightly barrel-shaped. The flowers are white.

DISTRIBUTION & HABITAT : Occurs on south-eastern Eyre Peninsula, including Boston Island, the northern part of Kangaroo Island, throughout the Mount Lofty Ranges (excluding the wettest parts) extending to the southern Flinders Ranges as far north as the Quorn area. Scattered populations also occur in the upper South-east, especially around Bordertown. It extends just over the border into Victoria. In the Mt Lofty Ranges it occurs mainly on the rocky, drier eastern slopes.

CULTIVATION & USES : Occasionally cultivated, it has a moderate growth rate and would be useful for shade, shelter and erosion control. Flowers are plentiful on good forms and it would be useful for broadscale planting such as along roadsides.

NOTES : Often confused with related box eucalypts such as *E. microcarpa, E. viridis* and *E. porosa*. Some populations of *E. odorata* in the south-east of the State and in Victoria have been called a new species, *E. silvestris*, however, it appears to be identical to *E. odorata* and thus is included in *E. odorata* here.

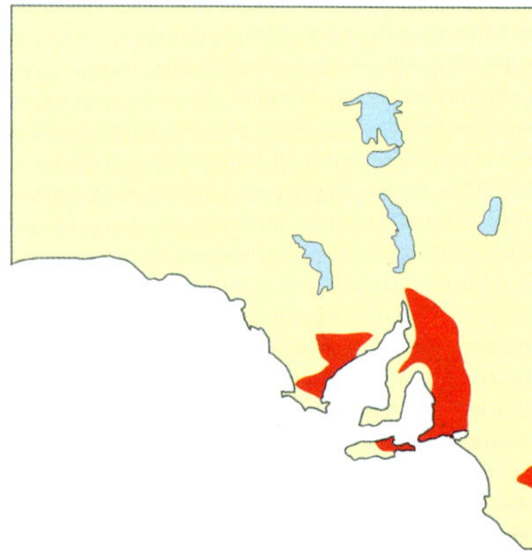

E. odorata on the Cummins to Yallunda Flat road, Eyre Peninsula, S.A.

Eucalyptus albopurpurea (Boomsma) Nicolle ined.

Purple-flowered Mallee; Port Lincoln Mallee

Genus equivalent: *Symphyomyrtus*; section: *Adnataria*; series: *Subbuxeales*

NAME : From the Latin *albus* meaning 'white' and *purpureus* meaning 'purple', referring to the range of flower colours.

DESCRIPTION : Varies from a dense mallee to 4 metres tall on Eyre Peninsula to a large mallee or scraggy tree to 12 metres tall on Kangaroo Island. The bark is usually rough at the base to about 2 metres up, then smooth olive-grey to coppery-pink and ribbony above. Adult leaves are glossy and green (the new growth may be dull). Buds and fruits are in umbels of seven to eleven that may appear in terminal clusters or along leafless sections of twigs. Fruits may be angular towards the base of fruit. The flowers are most often purple or pink, less often white on Eyre Peninsula, but on Kangaroo Island they are usually white and rarely pink.

DISTRIBUTION & HABITAT : Known from the southern side of Kangaroo Island and the southern tip of Eyre Peninsula. It occurs on sands or loams usually very close to the coast or in saline sites in mallee scrub with a variety of other mallees. Associated species include *E. diversifolia, E. rugosa, E. oleosa,* and *E. cosmophylla*

CULTIVATION & USES : Commonly cultivated with good forms being very ornamental with their masses of purple flowers. It is suitable for windbreaks and screening, even in exposed coastal situations on limestone soils. It prefers a well drained soil in a sunny position. Plants have a moderate growth rate and flower when quite young.

NOTES : The only eucalypt that commonly has purple flowers. It is distinguished from *E. lansdowneana* in the larger habit, shorter and broader leaves, non-red flowers and smaller buds and fruits. Closely related to *E. odorata*, and intergrades with it on central eastern Kangaroo Island, *E. albopurpurea* differs in the broader juvenile leaves; more glossy, broader adult leaves; often coloured flowers and larger buds and fruits. Most of the plants seen in cultivation are of the Eyre Peninsula form, with a low, bushy habit and commonly purple flowers.

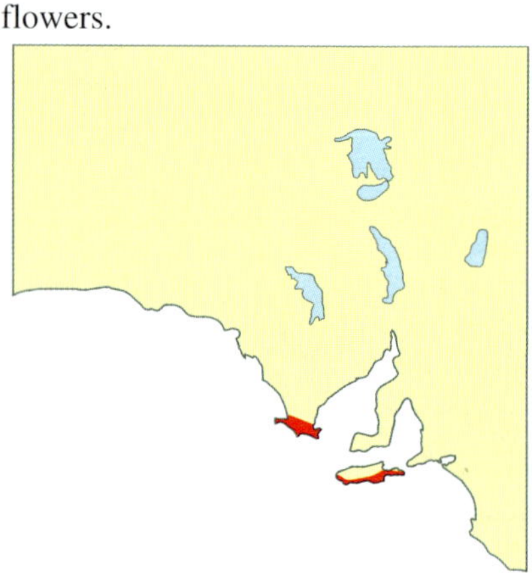

E. albopurpurea near Sleaford Mere, Eyre Peninsula, S.A.

Eucalyptus viridis R.T. Baker subsp. *viridis*

Green Mallee

Genus equivalent: *Symphyomyrtus*; section: *Adnataria*; series: *Subbuxeales*

NAME : From the Latin *viridis* meaning 'green', referring to the dark green adult leaves.

DESCRIPTION : A mallee, 4 to 7 metres tall. The bark is usually rough for 1 to 3 metres, grey-brown, then smooth above, tan to cream. The juvenile leaves are long and narrow, dark green. Adult leaves are glossy and green (immature adult leaves may be slightly blue-green at first before turning dark green). The buds and fruits occur in umbels of seven. The flowers are white.

DISTRIBUTION & HABITAT : Rare in South Australia, known from a few scattered, disjunct localities such as in the Flinders Ranges between Wyaka Bluff north of Quorn south to the Black Range, especially in the Alligator Gorge area. Also in the southern Middleback Range on northern Eyre Peninsula, such as on Iron Dutchess where it grows with *E. oleosa* and *E. gracilis*. *E. viridis* grows on rocky hill slopes and ridges in both the Flinders and Middleback Ranges. It is more common on the western plains of New South Wales, extending into southern Queensland and Victoria.

CULTIVATION & USES : Occasionally cultivated, it forms a tall, mallee or slender, several-stemmed tree in Adelaide. This habit and the narrow, dark green leaves make *E. viridis* subsp. *viridis* an attractive eucalypt for larger gardens and parks. In Victoria, eucalyptus oil is extracted from the leaves.

NOTES : Closely related to *E. odorata*, differing in the consistently mallee habit; the narrower juvenile and adult leaves; the smaller buds and fruits and its occurrence on rocky hill sides and ridges. *E. viridis* subsp. *wimmerensis* appears to be intermediate in characteristics between *E. viridis* subsp. *viridis* and *E. odorata*, but subsp. *viridis* always has narrower juvenile leaves than subsp. *wimmerensis*, and the two species occur in separate areas in South Australia.

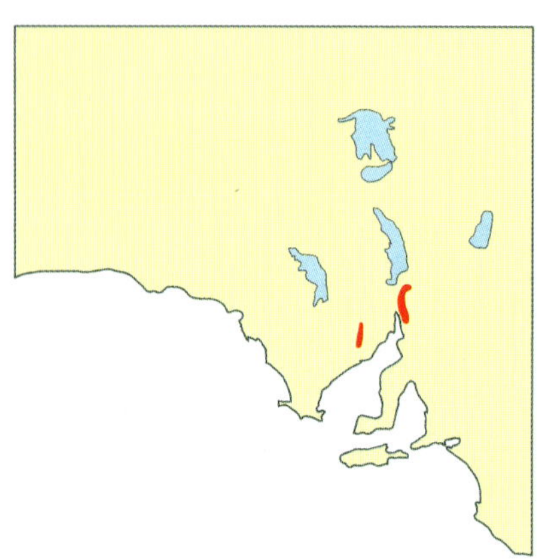

E. viridis subsp. *viridis* on Iron Duchess, Middleback Range, Eyre Peninsula, S.A.

Eucalyptus viridis R.T. Baker subsp. *wimmerensis* (K. Rule) Brooker & Slee

Wimmera Mallee

Genus equivalent: *Symphyomyrtus*; section: *Adnataria*; series: *Subbuxeales*

NAME : The subspecific name comes from the Wimmera region in Victoria, where the subspecies is most prevalent.

DESCRIPTION : A mallee with a tall, erect habit, 6 to 10 metres tall. The bark is rough for 1 to 2 metres at the base, thin and grey, then smooth thereafter, grey to orange. The juvenile leaves are glossy and green, similar but smaller than those of *E. odorata*. The adult leaves are narrow, glossy and dark green. The buds and fruits are almost identical to *E. viridis* subsp. *viridis*. The flowers are white.

DISTRIBUTION & HABITAT : Occurs in the Wimmera region of western Victoria, extending just into South Australia north and north-west of Bordertown. It grows in sands or loams in flat areas or on slight rises in mallee scrub, associated with species such as *E. arenacea, E. leptophylla, E. incrassata* and *E. behriana*.

CULTIVATION & USES : An attractive large mallee, sometimes developing into a small, erect tree in cultivation, although it is rarely cultivated. Plants have a moderate growth rate. It may prove useful for shade, shelter and woodlots in dry, southern areas of the State.

NOTES : This species appears intermediate in characteristics between *E. odorata* and *E. viridis* subsp. *viridis*. It grows near *E. odorata* around Bordertown but can be distinguished by its consistent mallee habit; the rough bark for up to 2 metres only and the slightly narrower adult leaves. From *E. viridis* subsp. *viridis* it differs in the shorter, broader juvenile leaves and the slightly larger adult leaves. All three differ from *E. microcarpa,* which also grows around Bordertown, in their smaller habit; non-terminal buds and fruits and narrow juvenile leaves (larger, broad juvenile leaves in *E. microcarpa*).

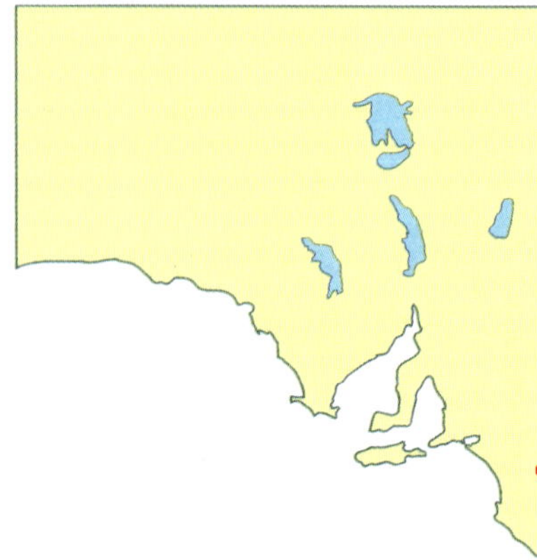

E. viridis subsp. *wimmerensis* on the Bordertown to Pinnaroo road, north-west of Bordertown, S.A.

Eucalyptus species 'Flinders Ranges'

Flinders Range Box

Genus equivalent: *Symphyomyrtus*; section: *Adnataria*; series: *Subbuxeales*

NAME : An undescribed and little-known species endemic to the Flinders Ranges, previously included in *E. viridis*.

DESCRIPTION : A mallee, of sprawling or upright habit, 2 to 6 metres tall and often thin-stemmed. The bark is smooth, ribbony, coppery-tan to grey throughout, or a little loose rough bark may be present at the base. The adult leaves are dull to rarely slightly glossy, blue-green. The small buds and fruits are often glaucous and occur in umbels of seven to eleven. The flowers are white.

DISTRIBUTION & HABITAT : Restricted to the northern Flinders Ranges of South Australia from about Mount Arden in the south to the Gammon Ranges in the north. It occurs on the slopes and high ridges of the ranges in gravelly loams or very rocky soils, sometimes with *E. flindersii, E. intertexta* or *E. porosa.* On parts of the Gammon Plateau, it forms dense, inpeneratable stands of thin-stemmed mallees codominant with *Melaleuca uncinata.* Endemic to South Australia.

CULTIVATION & USES : Only very recently tried in cultivation, it grows rapidly like the commonly cultivated *E. viridis*. Growing conditions would be the same as *E. viridis.* With its bluish crown of narrow leaves, it should prove to be quite ornamental.

NOTES : Closely related to *E. viridis*, differing in the usually completely smooth bark; dull, blue-green leaves and often glaucous buds and fruits. *E.* sp. 'Flinders Ranges' is far more common in the Flinders Ranges than *E. viridis*. Also closely related to *E. polybractea*, from which it differs in the smoother bark and narrower juvenile and adult leaves. Mallees intermediate between *E.* sp. 'Flinders Ranges' and *E. odorata* occur around Devils Peak and groups of mallees known as *E. 'desquamata'* at the base of this mountain may be a hybrid between these and *E. percostata*, which also grows there. *E.* sp. 'Flinders Ranges' differs from *E. odorata* in the smaller, mallee habit; smoother bark; narrower juvenile and adult leaves and the often glaucous buds and fruits.

E. species 'Flinders Ranges' on the slopes of the Elder Range, northern Flinders Ranges, S.A.

Eucalyptus polybractea R.T. Baker

Blue-leaved Mallee

Genus equivalent: *Symphyomyrtus*; section: *Adnataria*; series: *Subbuxeales*

NAME : From the Greek *poly* meaning 'many' and *bractea* meaning 'bract', referring to the numerous bracts around the buds at an early stage of their development, common to many eucalypts.

DESCRIPTION : A mallee of poor form, 2 to 6 metres tall. The bark may be rough for up to a metre, fibrous and grey or completely smooth, cream to grey-brown. The twigs are usually glaucous. Adult leaves are dull, blue-green and often glaucous when new. The buds are glaucous and occur in umbels of seven to eleven. The fruits may be glaucous when new, but become non-glaucous with age due to weathering. The flowers are white.

DISTRIBUTION & HABITAT : Only known from a single population in South Australia in the Mount Lofty Ranges north-east of Adelaide. It occurs in the Para Wirra Recreation Park near the Devil's Nose walking trail on the side of a high ridge. Here it grows with and in places as an understorey to *E. leucoxylon* subsp. *leucoxylon* and *E. odorata* with *E. goniocalyx* growing nearby. It is common and more widespread in northern Victoria and central New South Wales.

CULTIVATION & USES : Not a particularly ornamental mallee and therefore uncommon in cultivation. It is grown in New South Wales for *Eucalyptus* oil production and when cut to ground level produces bushy new growth rapidly.

NOTES : The single population of this species in South Australia is secure within a recreation reserve but it is far from the nearest populations in Victoria. It is similar to *E. odorata*, with which it grows in S.A., differing mainly in the poorer form and glaucous parts, particularly the buds. It is possible that the Para Wirra population of *E. polybractea* is a form recently diverged from typical *E. odorata* and which now appears distinct from it and very similar to *E. polybractea* from Victoria and N.S.W. *E. polybractea* is similar to *E.* sp. 'Flinders Ranges', differing in the broader juvenile and adult leaves and the rough bark, especially on larger mallees.

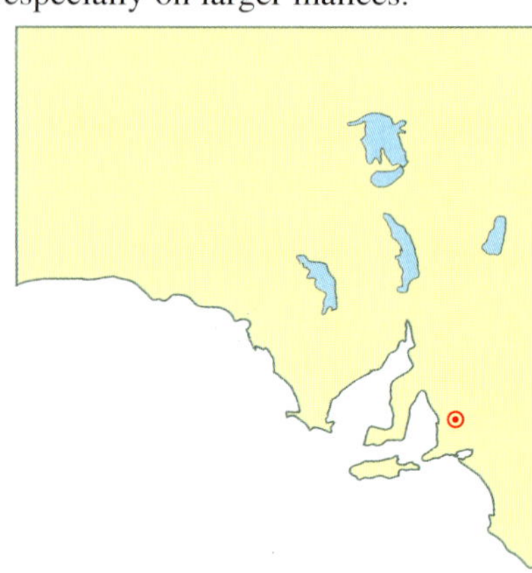

E. polybractea in Para Wirra Recreation Park, Mount Lofty Ranges, S.A.

Eucalyptus porosa F. Muell. ex Miq.

Mallee Box; South Australian Mallee Box; Quorn Mallee

Genus equivalent: *Symphyomyrtus*; section: *Adnataria*; series: *Subbuxeales*

NAME : From the Latin *porosus* meaning 'pierced with small holes', referring to the pores in the anthers, common to all the boxes (section *Adnataria*) as well as in other groups of eucalypts.

DESCRIPTION : Varies from a tree to 14 metres tall to a scraggy mallee on rocky bluffs, 2 to 8 metres tall. The bark is rough on the trunk and branches, light to dark grey, with smooth, light grey to coppery bark on the smaller branches. Small mallees may be mostly smooth-barked. Adult leaves are very variable in width, and are glossy, light green to green, with the intramarginal vein well in from the leaf edge. Buds and fruits are in umbels of seven. The flowers are white or very rarely deep red, occurring mostly in summer.

DISTRIBUTION & HABITAT : Occurs from north of Streaky Bay (including Eba Island) eastwards, including all of the Gawler Range (north to Lake Everard), Eyre and Yorke Peninsulas, in the Flinders Ranges as far north as the Gammon Ranges and south to the Keith area, but is absent from the wettest parts of the Mount Lofty Ranges. It extends into adjacent parts of New South Wales and Victoria. It usually occurs in heavy soils over limestone in large shallow depressions, but sometimes grows on rocky slopes such as in parts of the Flinders and Middleback Ranges.

CULTIVATION & USES : Commonly cultivated, suitable for any soil from sands to heavy clay, including soils over shallow limestone. An attractive tree in cultivation, suitable for broad-scale plantings (eg. farms and road-side). An excellent shade tree for areas of low to moderate rainfall. Plants are very quick growing when young and attract few pests and diseases.

NOTES : Commonly confused with other box species such as *E. odorata* and *E. behriana*, it differs in the more glossy, lighter green adult leaves with the intramarginal vein distinct and well in from leaf edge and the larger, more rounded fruit. It usually has broader leaves than *E. odorata*. On coastal cliffs at Newland Head, *E. porosa* is reduced to a sprawling mallee under one metre tall.

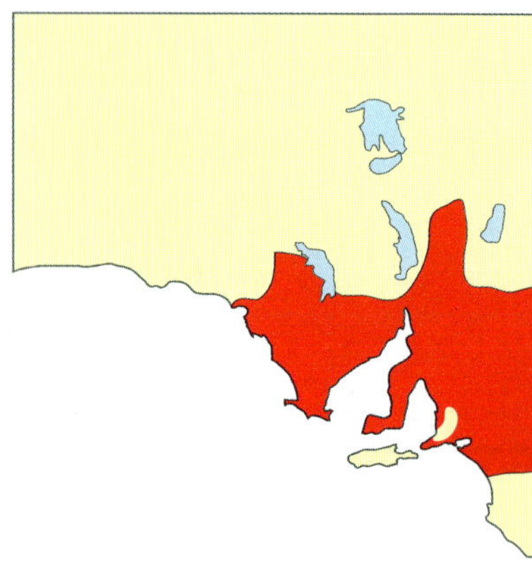

E. porosa on the Dukes Highway near Tailem Bend, S.A.

Eucalyptus fasciculosa F. Muell.

Pink Gum; Scrub Gum; Hill Gum

Genus equivalent: *Symphyomyrtus*; section: *Adnataria*; series: *Heterophloiae*

NAME : From the Latin *fasciculus* meaning 'fascicle' referring to the flower structure of this species. The common name refers to the dark pink colour of the very hard heartwood.

DESCRIPTION : A tree, often of poor form, 5 to 18 metres tall, rarely stunted to a mallee on very rocky sites. The bark is mostly smooth and patchy, cream and grey, often with some loose rough bark near the base. The twigs are glaucous in the south-eastern forms and sometimes on Kangaroo Island. The leaves are dull to slightly glossy and green. The buds and fruits are in umbels of seven which form larger terminal clusters. The operculum (cap) of the buds is usually narrower than the hypanthium (base). Fruits have a thin rim that is sometimes split. The flowers are white, occurring at any time of the year, especially autumn.

DISTRIBUTION & HABITAT : Common in the Mount Lofty Ranges, extending to the Coorong and eastward just into Victoria. Also on Kangaroo Island. It inhabits soils of poor fertility, usually sandy or rocky areas, often associated with *E. leucoxylon* and *E. odorata* in the Mount Lofty Ranges and the South-east and *E. odorata* or *E. cladocalyx* on Kangaroo Island.

CULTIVATION & USES : Rarely cultivated but common in the hills and coastal regions south of Adelaide. Plants have a moderate growth rate. It is suited to any well drained soil and is useful for very sandy, infertile soils. The fairly open crown allows other plants or grasses to grow underneath the tree. An attractive tree that usually has a crooked trunk and branches and although the individual flowers are tiny, they are often conspicuous because they occur in large clusters. A useful honey and woodlot tree.

NOTES : Trees look very similar to *E. leucoxylon* from a distance, but the much smaller buds and fruits in terminal clusters of *E. fasciculosa* distinguish it. Distinguished from the previous box (section *Adnataria*) species by the patchily smooth, mottled bark.

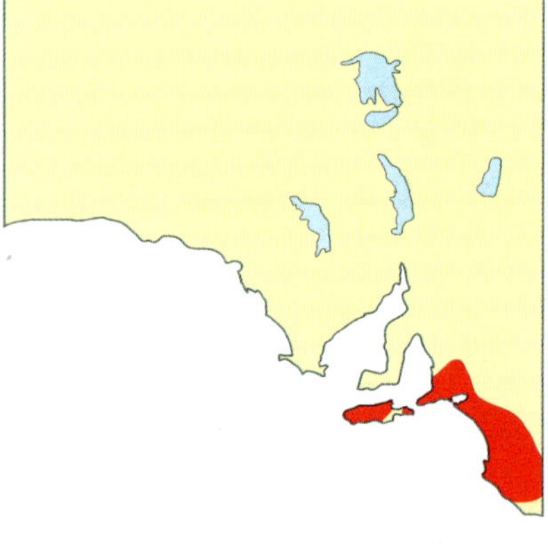

E. fasciculosa south-west of American River, Kangaroo Island, S.A.

E. fasciculosa near Currency Creek (looking towards Coorong), Fleurieu Peninsula, S.A.

Eucalyptus leucoxylon F. Muell. subsp. *stephaniae* K. Rule

Scrubby Blue Gum

Genus equivalent: *Symphyomyrtus*; section: *Adnataria*; series: *Melliodorae*

NAME : This subspecies name is dedicated to the author's daughter, Stephanie, who the author states 'was a constant and enthusiastic companion during *Eucalyptus* field trips and who died suddenly and unexpectedly in October, 1986'.

DESCRIPTION : A tree of poor form, reduced to a mallee on poor sites, 4 to 15 metres tall. The bark is mostly smooth, mottled cream, grey, olive green and tan, or on larger trees rough at the base or over the lower half of trunk, coarsely fibrous, grey, then smooth above. All parts are non-glaucous. The juvenile leaves are sessile. The adult leaves are glossy and light green to green. Buds and fruits occur in umbels of three. The fruits are cupular, 6 to 8 mm long by 7 to 11 mm wide, on short distinct pedicels. The flowers are white to rarely pink.

DISTRIBUTION & HABITAT : Occurs from south of Lake Albert near Meningie north-east to north of Coomandook and south-east to Salt Creek and eastwards from these locations into adjacent Victoria. It occurs in scattered stands on a variety of sites. Around Meningie and at Salt Creek it grows as a mallee in sands and loams in mallee scrub with *E. diversifolia* and *E. rugosa*, and north of Bordertown it grows in heavier soils between broad sandridges in pure stands.

CULTIVATION & USES : This subspecies is rare in cultivation. It is tolerant of shallow limestone and is often found in sandy soils. It may be useful for shelter and shade, although its form is generally poor compared to the other subspecies of *E. leucoxylon*.

NOTES : This subspecies can be distinguished by its small cupular fruits on short pedicels and its lack of glaucousness on the juvenile and adult material. Unlike *E. leucoxylon* subsp. *leucoxylon* and subsp. *megalocarpa*, the flowers are very rarely coloured.

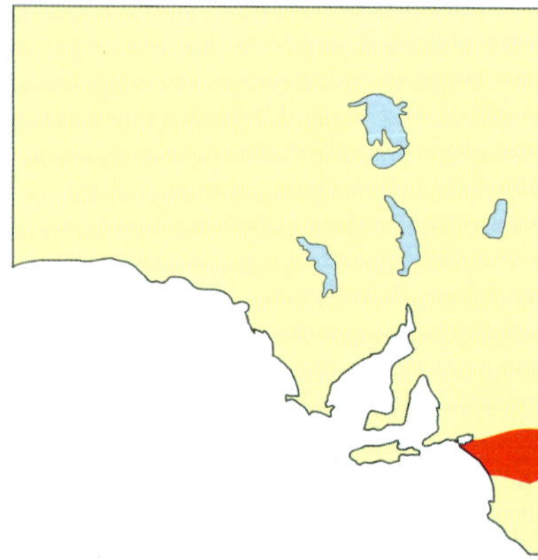

E. leucoxylon subsp. *stephaniae* on the Bordertown to Pinnaroo road, S.A.

Eucalyptus leucoxylon F. Muell. subsp. *pruinosa* (F. Muell. ex Miq.) Boland

Blue Gum; Yellow Gum (Vic)

Genus equivalent: *Symphyomyrtus*; section: *Adnataria*; series: *Melliodorae*

NAME : The subspecies name is from the Latin *pruinosus* meaning 'pruinose' (covered with a waxy, white powder), referring to the glaucous juvenile leaves, and sometimes the twigs, buds and fruits of this subspecies.

DESCRIPTION : A tree 6 to 20 metres tall. The bark is mostly smooth, flaky, yellow-cream to grey. A little rough, flaky bark may be present on the lower trunk. The twigs are often glaucous. The juvenile leaves are sessile, opposite and glaucous. The adult leaves are dull to slightly glossy, green to grey-green, sometimes glaucous. The buds and fruits are in groups of three on short pedicels and are often glaucous. The fruits are 6 to 9 mm long by 7 to 12 mm wide. The flowers are cream.

DISTRIBUTION & HABITAT : Occurs in two areas in South Australia. Firstly in the State's South-east, from north of Bordertown southwards and west to near Kingston S.E. where it grows in heavier soils in woodland in level areas. It also extends into Victoria as far east as Melbourne. The second area is in the northern Mount Lofty Ranges from the Barossa Valley area north to the Duchmans Stern in the southern Flinders Ranges where it grows on the undulating hills including mountain tops (e.g. Mt. Remarkable), again preferring heavy soils.

CULTIVATION & USES : Uncommon in cultivation but should be planted more, being quite ornamental. It would be useful for shade and woodlots. It prefers heavier soils than *E. leucoxylon* subsp. *stephaniae*, which grows in a similar area in the south-east of the State.

NOTES : This subspecies can be distinguished from the other subspecies of *E. leucoxylon* in the glaucous juvenile leaves and the often glaucous twigs, buds and fruits, sometimes giving the crown a bluish look. The other subspecies of *E. leucoxylon* are never glaucous. A form of this subspecies from near Kingston S.E. has larger leaves, buds and fruits than typical (fruits up to 12 mm long by 14 mm wide). In The Barrosa Valley area, *E. leucoxylon* subsp. *leucoxylon* and subsp. *pruinosa* intergrade.

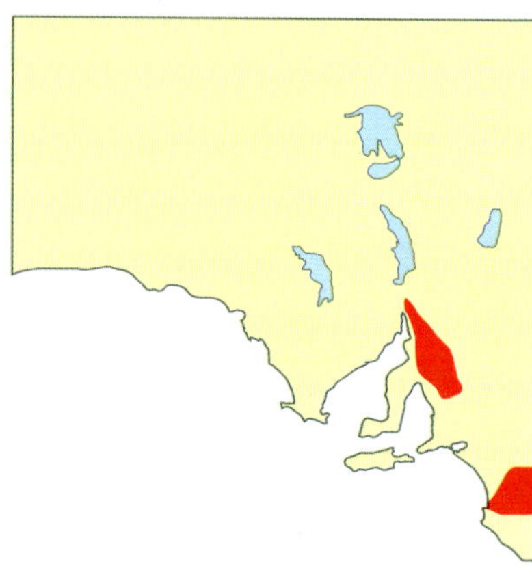

E. leucoxylon subsp. *pruinosa* north of Bordertown, S.A.

juvenile leaves

Eucalyptus leucoxylon F. Muell. subsp. *leucoxylon*

South Australian Blue Gum; White Ironbark; Yellow Gum (Vic)

Genus equivalent: *Symphyomyrtus*; section: *Adnataria*; series: *Melliodorae*

NAME : From the Greek *leucos* meaning 'white' and *xylon* meaning 'wood', referring to the pale-coloured timber of this species.

DESCRIPTION : A tall tree up to 30 metres tall or on rocky slopes reduced to a mallee down to 3 metres tall. The bark is mostly smooth, cream to grey or tan, although some rough, loose bark is sometimes present on the lower trunk. The juvenile leaves are sessile and opposite. Adult leaves are slightly glossy and green. The buds and fruits are in pendulous umbels of three. The fruits are wine glass-shaped and 9 to 14 mm long by 7 to 10 mm wide. The flowers range in colour from cream through pink to deep red.

DISTRIBUTION & HABITAT : Occurs on Kangaroo Island, most of the Mount Lofty Ranges as far east as near Murray Bridge, with some scattered occurrances in the southern Flinders Ranges (Wirrabara area). Also south of the Grampians in western Victoria and just extending into South Australia near Penola.

CULTIVATION & USES : Commonly cultivated, although less so than the similar *E. petiolaris* and *E. leucoxylon* subsp. *megalocarpa*. It attains a greater size than related species and is suitable for shade, timber (woodlots) and honey production because of its long flowering period. It is generally too large for gardens and as a street tree, where *E. petiolaris* is preferred. It grows well in any soil of reasonable drainage in areas of moderate rainfall.

NOTES : Easily confused with *E. petiolaris*, differing in the sessile juvenile leaves and usually somewhat smaller fruits. From *E. leucoxylon* subsp. *stephaniae* it differs in its larger habit, longer peduncles and pedicels and the larger buds and fruits, from subsp. *pruinosa* in the non-glaucous juvenile and adult material and the longer peduncles and pedicels; the often coloured flowers and larger buds and fruits and from subsp. *megalocarpa* in the taller, less dense habit, the narrower adult leaves and smaller buds and fruits.

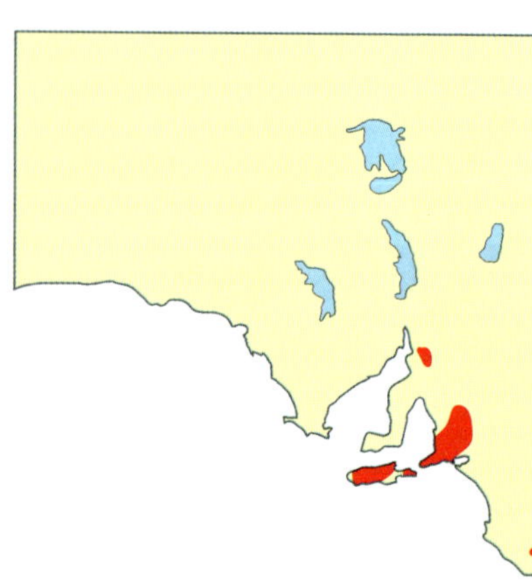

E. leucoxylon subsp. *leucoxylon* near Scott Conservation Park, Fleurieu Peninsula, S.A.

Eucalyptus leucoxylon F. Muell. subsp. *megalocarpa* Boland

Large-fruited South Australian Blue Gum

Genus equivalent: *Symphyomyrtus*; section: *Adnataria*; series: *Melliodorae*

NAME : From the Greek *megalo* meaning 'big' and *carpos* meaning 'fruit', referring to the large fruits of this subspecies.

DESCRIPTION : A spreading tree with a dense crown, 5 to 10 metres tall, reduced to a near prostrate shrub when exposed to strong salt spray. The bark is mostly smooth, flaky, cream to grey and tan, or sometimes rough, thin and flaky over the lower trunk. The juvenile leaves are sessile and non-glaucous. The adult leaves are glossy and green to dark green. The pendulous buds and fruits occur in umbels of three on long peduncles and pedicels (both to 20 mm long). The fruits are 12 to 18 mm long by 9 to 14 mm wide. The flowers are commonly white to less commonly pink to rarely dark red.

DISTRIBUTION & HABITAT : Of scattered coastal distribution in the south-east of the State, mainly from about Robe southwards, extending just into Victoria. It is the dominant tree around Port Macdonnell. It grows in sands and loams often over limestone. It may be the dominant eucalypt or may be associated with *E. diversifolia* on limestone rises or with *E. ovata* in depressions.

CULTIVATION & USES : Commonly cultivated, it forms and attractive densely-crowned tree suitable for shade, shelter and for ornamental purposes. Pink and red-flowered forms are more common in cultivation than in the wild. It is a better coastal species than the other subspecies of *E. leucoxylon*. In cultivation, the often confused *E. petiolaris* has a more crooked habit and more open crown than *E. leucoxylon* subsp. *megalocarpa*.

NOTES : Distinguished from the other *E. leucoxylon* subspecies by the coastal distribution; the broader adult leaves giving a denser crown and the larger buds and fruits on long pedicels. Sometimes confused with *E. petiolaris*, especially in cultivation, differing in the sessile juvenile leaves and generally broader adult leaves giving a denser crown.

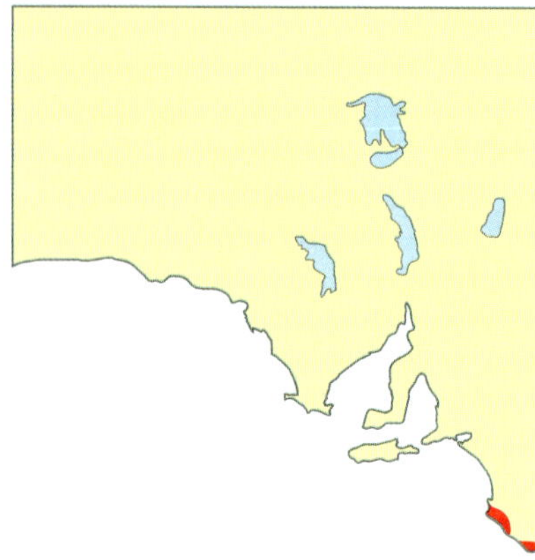

E. leucoxylon subsp. *megalocarpa* in Clark Picnic Ground, near Port Macdonnell, S.A.

Eucalyptus petiolaris (Boland) K. Rule

Eyre Peninsula Blue Gum; Water Gum

Genus equivalent: *Symphyomyrtus*; section: *Adnataria*; series: *Melliodorae*

NAME : From the Latin *petiolaris* meaning 'petiolate', referring to the petiolate juvenile leaves in contrast to the closely related *E. leucoxylon*.

DESCRIPTION : Usually forming an attractive tree to 15 metres tall, although it becomes stunted and mallee-like on rocky or swampy sites. The bark is mostly smooth and mottled grey and cream although a little rough shaggy bark may occur near the base. The juvenile leaves are petiolate (have stalks). The adult leaves are slightly glossy and green. The buds and fruits occur in groups of three and may be slightly ribbed. The flowers are quite conspicuous and vary from commonly white through pink to red or occasionally yellow or apricot.

DISTRIBUTION & HABITAT : Restricted to Eyre Peninsula from Darke Peak to Port Lincoln. It grows in hilly areas often along creeks and is common in the Cleve and Koppio Hills where it grows in woodland, often associated with *E. cladocalyx* and *E. odorata*.

CULTIVATION & USES : Very commonly cultivated, this species is often sold as *E. leucoxylon* 'Rosea'. Although flower colour is not true to seed, flowers are often pink to red, both in the wild and in cultivation and the colourful flowers along with its graceful habit make it ideal for streets, large gardens and parks. The flowers are bird attracting. It is quick growing and suited to most soils and is one of the most salt tolerant eucalypts, being exceeded by only a handful of Western Australian wheatbelt species.

NOTES : This species is easily confused with *E. leucoxylon subsp. megalocarpa*, and has similar sized and shaped buds and fruits. *E. petiolaris* differs in its petiolate juvenile leaves and narrower adult leaves forming a more graceful crown. It has larger buds and fruits than the other subspecies of *E. leucoxylon*.

E. petiolaris on the Cummins to Yallunda Flat road, Eyre Peninsula, S.A.

Eucalyptus macrorhyncha F. Muell. ex Benth. subsp. *macrorhyncha*

Red Stringybark

Genus *Eucalyptus*; section: unnamed 'Eastern'; series: *Pachyphloiae*

NAME : From the Greek *macros* meaning 'large' and *rhynchos* meaning 'snout or beak', referring to the beaked operculum (bud cap).

DESCRIPTION : A tree 8 to 18 metres tall. The bark is rough throughout, thick, coarsely stringy and grey-brown to red-brown. The juvenile leaves are rough-hairy for a few pairs before becoming smooth. The adult leaves are glossy and green. The buds and fruits occur in umbels of seven to fifteen. The operculum of the bud is notably beaked. The new fruits have a conspicuous red disc and valves. The flowers are white.

DISTRIBUTION & HABITAT : Of very restricted distribution in South Australia, known only from south of Clare, in the hills west of Sevenhill, including Spring Gully Conservation Park. Here it grows in gravelly loams on hillsides often with *E. goniocalyx* subsp. *goniocalyx* and *E. leucoxylon* subsp. *pruinosa*. It is very common and widespread in the Great Dividing Range of Victoria and New South Wales.

CULTIVATION & USES : Rarely cultivated in South Australia. Plants have a slow to moderate growth rate. It will grow in most well-drained soil, except those with limestone. Red stringybark forms a broad crowned shady tree when grown in the open. It could be grown in woodlots for timber production.

NOTES : The thick, stringy bark, the glossy, green adult leaves and the strongly beaked cap to the buds are distinctive. This species and also *E. albens, E. bicostata, E. dalrympleana* subsp. *dalrympleana* and *E. viminalis* subsp. *viminalis* are remarkable in that their small, restricted South Australian populations are so far disjunct from their major eastern states distribution.

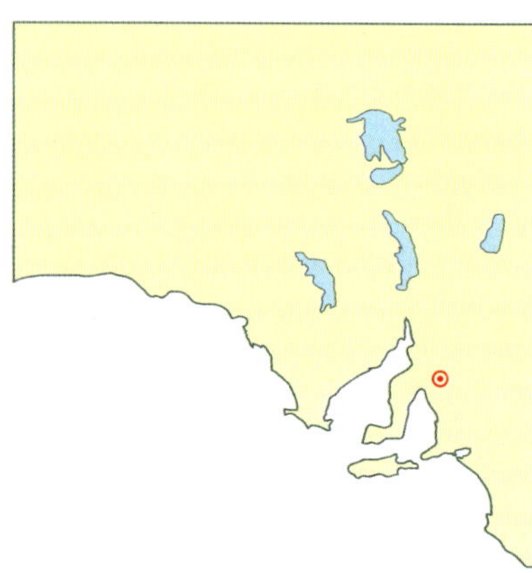

E. macrorhyncha subsp. *macrorhyncha* west of Sevenhill, S.A.

Eucalyptus arenacea Marginson & Ladiges

Sand Stringybark; Desert Stringybark; Brown Stringybark

Genus *Eucalyptus*; section: unnamed 'Eastern'; series: *Pachyphloiae*

NAME : From the Latin *arenaceus* meaning 'of sand' referring to the species occurrence on sandy soils.

DESCRIPTION : A large, robust mallee or multi-stemmed scraggy tree, 3 to 6 metres tall. The bark is rough throughout, coarsely stringy and grey-brown. The juvenile leaves are hairy-rough for at least 10 pairs and then smooth. The adult leaves are firm, 90 to 130 mm long by 20 to 35 mm wide, glossy and green. The buds and fruits occur in umbels of seven to fifteen. The buds have a smooth operculum. The fruits are 7 to 9 mm long by 9 to 11 mm wide, with a broad level to raised disc and usually with three valves. The flowers are white.

DISTRIBUTION & HABITAT : Occurs in the upper South-east of South Australia, from north of Kingston S.E. eastwards, being common in Ngarkat Conservation Park and north and south of Bordertown. It extends into adjacent Victoria as far as the northern Grampians. It always grows in poor white sands, usually on broad sand ridges and rises. It may be the dominant mallee in such sandy soils or may be associated with *E. leptophylla* and *E. incrassata*.

CULTIVATION & USES : Rare in cultivation, it is not known if the low mallee form will be retained if grown in more fertile soils. It requires a well drained soil in areas of moderate rainfall, where trials show it to be fairly quick growing. It would be very useful for shelter and soil erosion control in light soils.

NOTES : Distinctive because of its sandy habitat; the rough, stringy bark and the firm, glossy green leaves. It is closely related to *E. baxteri,* differing in its consistently poor, mallee-like habit; the juvenile leaves which are rough for more than 10 pairs; the narrower, tapering adult leaves; the smaller, non-warty buds and the smaller fruits with a less raised disc.

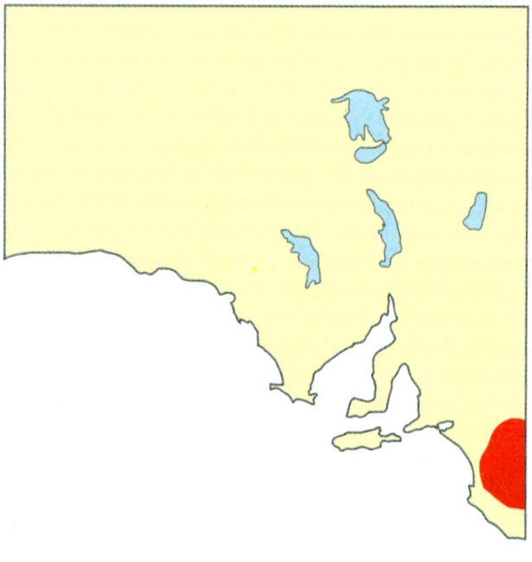

E. arenacea in Ngarkat Conservation Park, S.A.

Eucalyptus baxteri (Benth) Maiden & Blakely ex J.M. Black

Brown Stringybark

Genus *Eucalyptus*; section: unnamed 'Eastern'; series: *Pachyphloiae*

NAME : Named after W. Baxter, who collected the type specimens in 1828.

DESCRIPTION : A tree to 15 metres tall, often reduced to a mallee on exposed sites. The bark is rough throughout, coarsely stringy and grey-brown. The juvenile leaves are rough at first but become smooth before the eighth pair. Adult leaves are 55 to 120 mm long by 25 to 50 mm wide, glossy and green. The buds and fruits occur in umbels of seven to fifteen. The bud operculum is warty. The fruits are rounded with an ascending disc, usually with 4 valves. The flowers are white.

DISTRIBUTION & HABITAT : Restricted to higher rainfall areas such as in the Mount Lofty Ranges, especially on Fleurieu Peninsula and on western Kangaroo Island, and in the lower South-east in near coastal areas. It grows on fairly poor soils that are sandy, gravelly or occasionally somewhat clayey. It often grows with stunted *E. obliqua* as dense low woodland although it may be associated with a number of other species such as *E. fasciculosa, E. viminalis* subsp. *cygnetensis* and *E. cosmophylla*.

CULTIVATION & USES : Rarely cultivated, it is suitable for high rainfall areas in well drained soils. In exposed positions where it is bushy growing such as on ridges it is suitable for windbreaks and in better conditions it forms a shady tree suitable for parks and farms. Plants are fairly quick growing. It is intolerant of alkaline soils (limestone) and prolonged waterlogging.

NOTES : It can be confused with the very closely related *E. arenacea* and also *E. obliqua*, two similar looking stringybarks (series *Pachyphloiae*). From *E. arenacea* it differs in its juvenile leaves which become smooth before pair eight; the broader adult leaves; the larger warty buds and the larger fruits with a more steeply raised disc. From *E. obliqua* it differs in its rough juvenile leaves; its warty buds and its rounded fruits with an ascending disc, on longer pedicels.

E. baxteri near Nangkita, Fleurieu Peninsula, S.A.

Eucalyptus obliqua L'Hérit.

Messmate Stringybark; Messmate

Genus *Eucalyptus*; section: unnamed 'Eastern'; series: *Eucalyptus*

NAME : From the Latin *obliquus* meaning 'oblique' referring to the oblique bases of the leaves, also common to other eucalypts.

DESCRIPTION : A well-formed straight tree to 40 metres tall near Mount Lofty, reduced to a shrubby mallee only 2 metres tall in exposed coastal sites such as on Kangaroo Island. The bark is rough throughout, stringy and grey-brown. The juvenile leaves are smooth. The adult leaves are glossy and green, usually with oblique (unevenly joined) leaf bases. The buds and fruits occur in umbels of eleven to nineteen. The slender buds are non-warty. The fruits are barrel-shaped, with a steeply descending disc and enclosed valves. The flowers are white.

DISTRIBUTION & HABITAT : Occurs in the higher rainfall areas of the State's South-east, from about Naracoorte east towards Kingston S.E. and southwards, also occurs in the Mount Lofty Ranges and on the western half of Kangaroo Island. In the central Mount Lofty Ranges it forms tall, often pure forests sometimes with *E. dalrympleana* subsp. *dalrympleana* or *E. viminalis*. Elsewhere in South Australia it ranges from a large tree in forest vegetation to a shrubby mallee in coastal mallee scrub, usually on well drained gravelly clays or sands. It also grows in Victoria and New South Wales, just extending into Queenland, and in Tasmania, where it commonly attains heights of over 80 metres.

CULTIVATION & USES : Occasionally cultivated, mostly for forestry purposes in the eastern states. It requires high rainfall and for best growth, fertile soils. In such conditions it is very fast growing. Suitable for timber production and as a shade tree.

NOTES : *E. obliqua* is the type species for the genus *Eucalyptus*. In the south-east of the State *E. obliqua* can be very large-fruited, and this variable form has been described as *E. obliqua* var. *megacarpa*. *E. obliqua* can be confused with *E. baxteri*, differing in the smooth juvenile leaves; the non-warty buds and the barrel-shaped fruits.

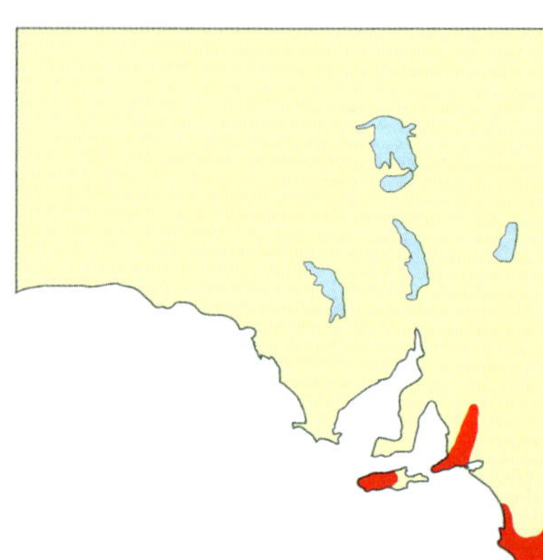

E. obliqua on the Willunga to Mount Compass road, S.A.

Eucalyptus willisii Ladiges, Humphries & Brooker subsp. *falciformis* Newnham, Ladiges & Whiffin

Peppermint

Genus *Eucalyptus*; section: unnamed 'Eastern'; series: *Radiatae*

NAME : The specific names honours Dr. J.H. Willis of Melbourne, who has made many notable contributions to the flora of Victoria. The subspecific name is from the Latin *falcatus* meaning 'sickle-shaped' or 'falcate' and *forma* meaning 'shape' referring to the prominently curved juvenile leaves

DESCRIPTION : A tree, sometimes multi-stemmed and mallee-like, 5 to 12 metres tall. The bark is rough on the trunk and usually the branches, finely fissured, grey, then smooth above. The adult leaves are glossy and green. The buds and fruits occur in umbels of eleven to fifteen. The flowers are white and occur in spring and summer.

DISTRIBUTION & HABITAT : It occurs in scattered populations in the South-east of the State, from Robe to Penola and southwards, extending into adjacent Victoria disjunctly east to the Otway Range, east of which it is replaced by *E. willisii* subsp. *willisii*. It occurs in sandy soils in broad, low areas, or sometimes on rises. Associated species include *E. camaldulensis, E. viminalis* subsp. *cygnetensis, E. pauciflora, E. baxteri* and *E. obliqua*.

CULTIVATION & USES : Rarely cultivated, trials show it to be very quick growing in sandy soils. The seed is easy but slow to germinate. It is suited to areas of assured rainfall in the southern part of the State.

NOTES : Formally known as *E. nitida*, a species restricted to Tasmania. The small flat-topped fruits of *E. willisii* are distinctive, although similar to *E. pauciflora*, from which it differs in its smaller habit; the rough bark; the narrower juvenile leaves; the smaller, thinner adult leaves with less parallel venation and the slightly smaller buds and fruits. *E. willisii* subsp. *falciformis* differs from subsp. *willisii*, an endemic to eastern Victoria and Flinders Island, in its larger juvenile and adult leaves and the larger fruits.

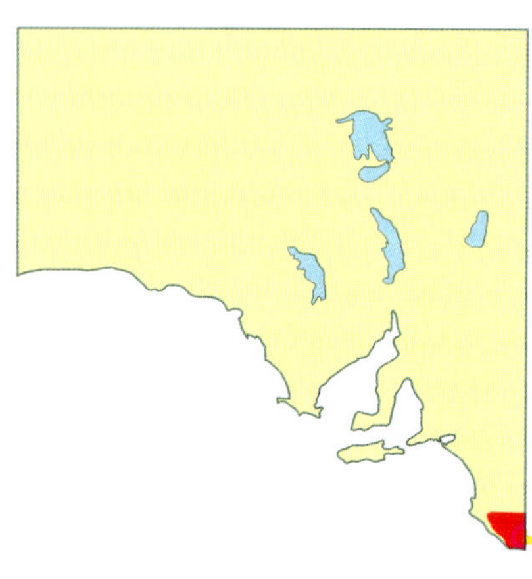

E. willisii subsp. *falciformis* east of Mount Richmond township, Victoria.

Eucalyptus pauciflora Sieb. ex Spreng. subsp. *pauciflora*

Snow Gum; Cabbage Gum

Genus *Eucalyptus*; section: unnamed 'Eastern'; series: *Pauciflorae*

NAME : From the Latin *paucus* meaning 'few' and *florus* meaning 'flowered', referring to the paucity of flowering. The name is somewhat misleading, as this subspecies flowers quite abundantly.

DESCRIPTION : A tree 8 to 20 metres tall in South Australia. The bark is smooth, grey and tan to white, sometimes with a little rough, ribbony grey-brown bark at the base. Distinctive scribbles are often present on the smooth bark, but these have not been seen in the South Australian population. Adult leaves are glossy and green with almost parallel side venation. Buds and fruits occur in umbels of nine to fifteen. The flowers are white, occurring in spring and summer.

DISTRIBUTION & HABITAT : Very rare in South Australia, known from only one population south-east of Mt Gambier in the far south-east of the State. It occurs in a flat area of about 1.5 by 1 km known as 'Snow Gum Reserve', surrounded by pine plantations within Caroline Forest Reserve. Here the snow gums grow in brown sand scattered amongst *E. baxteri* and *E. viminalis* subsp. *cygnetensis*. It is far more common in Tasmania, Victoria and New South Wales, extending just into Queensland.

CULTIVATION & USES : Rarely cultivated in South Australia because most of the State is too dry and hot and plants suffer heavily over summer. Seed is best kept in a freezer for several weeks before planting for best germination. It would be suited to well drained soils in cold, wet areas and will tolerate frost and snow. It is susceptible to saline water and soils.

NOTES : The smooth bark and near parallel side venation distinguish this species from other South Australian eucalypts, especially the related *E. willisii* subsp. *falciformis*. While in South Australia *E. pauciflora* grows near sea level (about 40 metres altitude), in the eastern states it grows up to about 2000 metres, the highest altitude of any eucalypt in Australia. Other altitudinal subspecies exist in Victoria and New South Wales.

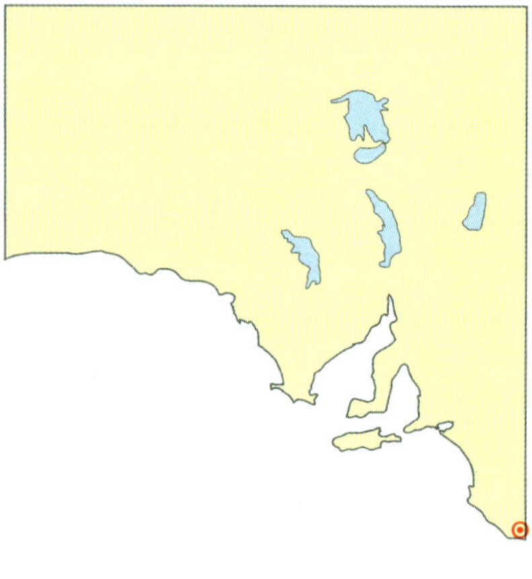

E. pauciflora subsp. *pauciflora* in Snow Gum Reserve, Caroline Forest Reserve, S.A.

Eucalyptus remota Blakely

Kangaroo Island Mallee Ash

Genus *Eucalyptus*; section: unnamed 'Eastern'; series: *Psathyroxyla*

NAME : From the Latin *remotus* meaning 'remote', referring to this species being remote from closely related species on the east coast of Australia.

DESCRIPTION : Varies from an erect-stemmed mallee to 5 metres tall in areas prone to fire, to a low branching tree to 15 metres tall near Cape Dutton. The bark is rough on the trunk and larger branches, finely furrowed and grey-brown, then smooth grey to yellow-cream thereafter. Small mallees may be completely smooth-barked. Juvenile leaves (conspicuous after fires) are up to 100 mm long by 50 mm wide, dull, grey-green and distinctly pendulous. Adult leaves are glossy and green to blue-green. Buds and fruits are in umbels of seven to thirteen. The fruits are obconical with a level disc (flat-topped). The flowers are white.

DISTRIBUTION & HABITAT : The only eucalypt to be naturally restricted to Kangaroo Island. It only grows on the western half of the island (west of about Mount Taylor) in gravelly sands and loams over laterite in mallee scrub, associated with species such as *E. baxteri, E. cosmophylla* and *E. diversifolia*. It is common in Flinders Chase National Park. Endemic to South Australia.

CULTIVATION & USES : Almost unknown in cultivation. Seed is slow to germinate, however plants are relatively quick growing in sandy soils. It requires a well-drained acidic soil (it does not tolerate limestone). It may be suitable for windbreaks and low shade in areas of moderate rainfall.

NOTES : Not closely related to any other eucalypts in South Australia, but often confused with *E. diversifolia*. *E. remota* differs in its petiolate and alternating juvenile leaves after pair four; the smaller buds and smaller, more obconical fruits; the longer pedicels and the consistently level disc of the fruit.

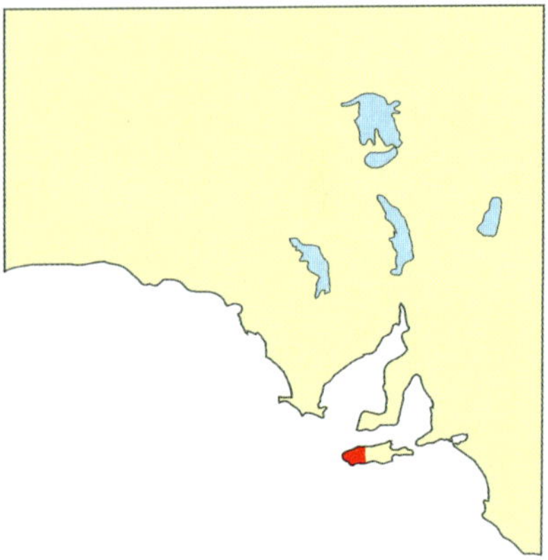

E. remota on eastern edge of Flinders Chase National Park, Kangaroo Island, S.A.

E. remota near Larrikin Lagoon, Flinders Chase National Park, Kangaroo Island, S.A.

dried fruit

Eucalyptus diversifolia Bonpl. subsp. *diversifolia*

Coastal Mallee; Coastal White Mallee; Soap Mallee

Genus *Eucalyptus*; section: unnamed 'Western'; series: *Diversiformes*

NAME : From the Latin *diversus* meaning 'diverse' and *folium* meaning 'leaf', probably referring to the difference between the juvenile and adult leaf stages.

DESCRIPTION : A mallee up to 10 metres tall, or prostrate and ground-hugging, under 50 cm tall on exposed coastal cliffs. The bark is usually completely smooth, grey to cream and somewhat ribbony. In large old mallees, some thin, rough, fibrous, grey bark may be present for up to 3 metres. Juvenile leaves are sessile and opposite, blue-green in colour. The adult leaves are glossy and green to blue-green. The buds and fruits are often prolific, occurring in umbels of seven to thirteen. The buds are diamond-shaped with a rounded to pointed operculum. The fruits are flat-topped or with a slightly ascending disc. The flowers are white.

DISTRIBUTION & HABITAT : Widely distributed in coastal areas of South Australia, over much of Eyre Peninsula inland to north of Minnipa, lower Yorke Peninsula, most of Kangaroo Island, around Newland Head with small populations near Carrickalinga and at Black Hill, and from from the Murray Lakes eastwards into the South-east, as far inland as north-west of Bordertown. It grows in sandy soils, always over limestone in mallee scrub. Endemic to South Australia.

CULTIVATION & USES : This species is very useful for exposed coastal situations where little else will grow. It has foliage to the ground in exposed situations and can be used for shelter and screening. It will not tolerate waterlogging or salt affected soils but is very resistant to salt spray and shallow limestone. Plants are generally quick growing.

NOTES : Very variable in habit depending on the environmental conditions in which it is growing. The diamond-shaped buds and large fruits with a broad, often flat disc are distinctive. May be confused with *E. remota* but differs in its opposite juvenile leaves; larger buds and fruits on shorter pedicels and usually cupular rather than obconical-shaped fruits.

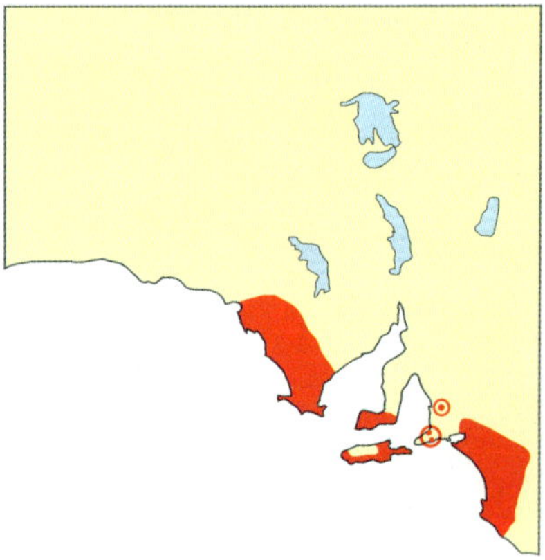

E. diversifolia subsp. *diversifolia* south of Lock, Eyre Peninsula, S.A.

Eucalyptus diversifolia Bonpl. subsp. *hesperia*

Wright & Ladiges

Wylie Scarp Mallee

Genus *Eucalyptus*; section: unnamed 'Western'; series: *Diversiformes*

NAME : The subspecific name is from the Latin *hesperis* meaning 'western' referring to the western occurence of this subspecies to subsp. *diversifolia*.

DESCRIPTION : A mallee, often of low, spreading habit, up to 4 metres tall, reduced to a groundcover above the Nullarbor cliffs. The bark is mostly smooth, grey to cream, often with some rough, stringy, grey bark near the base. The juvenile leaves are opposite and sessile and lanceolate rather than ovate-lanceolate in subsp. *diversifolia.* The adult leaves are dull to glossy and green. The buds occur in umbels of seven to thirteen. The fruits are cupular, with a broad, flat disc. The flowers are white.

DISTRIBUTION & HABITAT : Occurs along a coastal strip from just east of the W.A.-S.A. border in South Australia westwards to Toolina Cove in Western Australia. It occurs on limestone cliffs such as on the Wylie and Hampton Scarps or in shallow sandy soils over limestone above the Nullarbor cliffs where it is sometimes associated with *E. yalatensis* and *E. gracilis*.

CULTIVATION & USES : Not known in cultivation, this subspecies should have similar growth requirements as *E. diversifolia* subsp. *diversifolia*, and would have similar uses in difficult exposed sites in limestone soils.

NOTES : Very similar to *E. diversifolia* subsp. *diversifolia*, subsp. *hesperia* differs in the longer, narrower juvenile leaves; the smaller leaves and the smaller fruits with a broad, level disc. *E. diversifolia s*ubsp. *hesperia* is somewhat intermediate in characteristics between subsp. *diversifolia* and an undescribed eucalypt species from a few ironstone hills near Norseman in Western Australia.

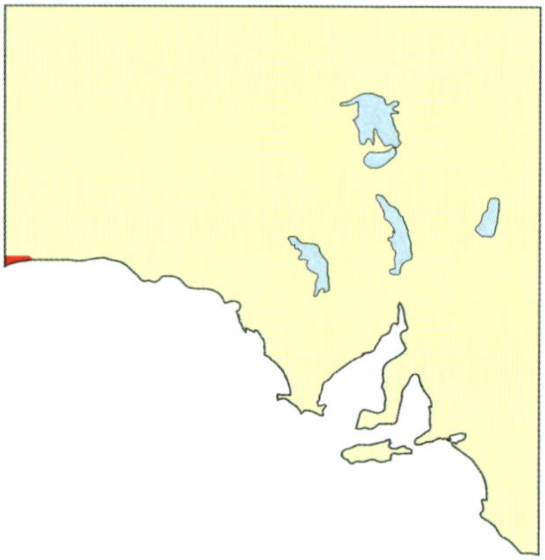

E. diversifolia subsp. *hesperia* east of Border Village, S.A., overlooking the Great Australian Bight.

Appendix - Eucalypt species as yet unrecorded in South Australia but occurring close to the S.A. border.

Corymbia aparrerinja - Small tree of Western Australia, Northern Territory and Queensland, growing as far south as the Petermann Ranges (N.T.) and Giles (W.A.). The common 'ghost gum' of central Australia, with smooth, powdery white bark; narrow, green leaves and small papery fruits that open soon after maturity.

C. opaca - Small tree of Western Australia and Northern Territory, occurring south to just north of the S.A.-N.T. border on the Stuart Highway. Related to *E. terminalis,* differing in the less coarse leaves, buds and fruits and the more slender peduncles and pedicels.

C. chippendalei - Small, sometimes several-stemmed tree from Western Australia and Northern Territory, growing close to S.A. in the Warburton to Yulara area. The 'sandhill bloodwood' restricted to the crests and slopes of sand ridges, closely related to *C. terminalis*, differing in the non-scurfy buds and the broader, more rounded fruits on short thick pedicels.

Eucalyptus oldfieldii - Mallee endemic to Western Australia occurring as far east as south of Warburton in the Great Victoria Desert. Related to *E. oxymitra*, differing in being consistently non-glaucous, and the sessile buds and fruits in umbels of three.

E. pachyphylla - Mallee of Western Australia, Northern Territory and Queensland, extending south to near the S.A. border near Giles (W.A). Closely related to *E. kingsmillii* (especially subsp. *kingsmillii*) differing in the much less ribbed buds and fruits on shorter peduncles and pedicels. The flowers are pale yellow.

E. kitsoniana - Mallee endemic to Victoria occurring as far west as the lower Glenelg River near Nelson. A mallee of section *Maidenaria*, with no close relatives. Distinguished by its combination of swampy habitat; mallee habit; smooth olive-green to tan bark; very thick, glossy, light green adult leaves and sessile buds and fruits in groups of seven.

E. populnea* subsp. *bimbil - Small tree of New South Wales and southern Queensland, growing close to the S.A. border north of Broken Hill. A distinctive box (section *Adnataria*) with rough, finely fissured bark throughout and broad, very thin, very glossy, green leaves on long petioles (stalks).

E. normantonensis - Mallee of Queensland, Northern Territory and Western Australia (Pollock Hills only), known no closer than the MacDonnell Ranges (N.T.) but of very scattered occurrence and possibly in S.A. Similar to *E. sparsa*, differing in the narrower, less glossy leaves, the smaller buds and fruits and the outer stamens of the flower lacking anthers (all fertile in *E. sparsa*).

E. lucasii - Mallee endemic to Western Australia, occurring as far east as near the Wanna Lakes, east of Neale Junction in the Great Victoria Desert. A box eucalypt (section *Adnataria*), similar in appearance to *E. intertexta*, differing in the smaller, mallee habit, the usually smooth bark and the outer stamens of the flower lacking anthers (all fertile in *E. intertexta*).

E. melliodora - Tree of Queensland, New South Wales and Victoria, occurring as far west as west of the Grampians (Victoria). Related to *E. leucoxylon*, differing in the buds and fruits which are in umbels of seven and the bark, which is variable, but often rough on the trunk and branches.

E. sideroxylon - Tree of Victoria, New South Wales and Queensland, not occurring particulary close to South Australia, however, a single tree of this species is know at Oulnina Park near Olary in S.A. and was most likely planted long ago. Related to *E. leucoxylon*, with similar buds and fruits in umbels of three, differing most strikingly in the rough, hard, deeply furrowed bark (ironbark) on the trunk and branches.

Glossary

adult (of the leaves) - The final growth phase of the leaves.
anther - the tiny, pollen bearing organ at the tip of the stamen.

conical - cone-shaped
connate (of the leaves) - when the leaf bases of opposite leaves are joined around the stem.
crenulate (of the leaves) - with a finely scalloped edge.
crown - The upper, leafy part of a plant.
cupular - cup-shaped.

decurrent (of the leaf) - when the midrib (central vein) joins lower down the stem than the leaf margins on a sessile leaf.
disc (of the fruit) - a circular band present on the fruit between the rim and the valves.
discolorous (of the leaves) - where the upper leaf surface is darker in colour than the lower surface.
disjunct - occurring in two or more distinct areas separated from one another by a large distance.

elliptical - more or less shaped like a flattened circle or oval.
endemic - occurring naturally only in the area defined.
exserted (of the valves) - projecting above the rim of the fruit.

genus equivalent - a taxonomic rank used here that is considered to be at equal rank to that of genus, but is currently a subgenus within *Eucalyptus*.
glaucous - covered with a white, waxy substance that can usually be rubbed off.

fruit - the 'gum nut', which is a swollen hypanthium, enclosing the seeds, which form following pollination of the flowers.

habit - the overall form and gross structure of the plant.
habitat - the environment in which the plant grows.
hypanthium - the basal part of the flower and bud which later swells to form the fruit.

intergrade - where two or more closely related species, subspecies or varieties are in geographical contact and appear to interbreed with each other.
intramarginal vein - the first prominent vein in from, and running parallel with, the leaf edge.

juvenile (of the leaves) - the growth phase of the leaves before adult leaves are produced, seen in the seedling stage and as regrowth from damaged plants. Here, juvenile leaves are used to include seedling leaves.

lanceolate - lance-shaped, that is long and relatively narrow and broadest near the base and tapering to a point.

laterite - red or orange, gravelly clay, the gravel usually being ironstone.

mallee - a growth form of many eucalypts, where many stems of about equal size arise from ground level from an individual plant.
minniritchi - a distinctive bark type where the old bark partially peels off in thin longitudinal strips that curl and remain attached to reveal the under-bark, usually of a lighter colour.

obconical - shaped like an upturned cone.
oblique (of the leaves) - where the two sides of the leaf base are joined to the petiole (stalk) at different points.
operculum - the upper part of the bud that falls off upon flowering to expose the flowers (stamens). Also known as the bud cap.
ovate (of the leaf) - egg-shaped and widest nearer the base (petiole) of the leaf.

pedicel - the individual stalk of the bud, flower or fruit, not present in some species.
peduncle - the common stalk of the cluster (umbel) of buds, flowers or fruits.
petiole - the leaf stalk.
pith - the inner core of a plant's stem.
provenanance - a particular place or area of origin.

scribbles - irregular, scribble-like markings on the bark caused by burrowing insects.
scurfy - with a finely rough surface.
sessile - lacking a stalk (of the leaf, lacking a petiole; of the bud, flower and fruit lacking a pedicel; of the umble, lacking a peduncle).
stamen - a thin filament surmounted by an anther, many of which form the conspicuous part of a eucalypt flower.
subsp. - abbreviation of subspecies (plural subspp.). Subspecies of a species have less distinctiveness to one another than species do to one another, and usually occupy a different geographical area to the other subspecies of that species.

terminal (of the buds and flowers) - occurring in usually large clusters at the ends of the branchlets; (of the crown) - with most leaves concentrated towards the ends of branches on the outside of the plant.
tessellated (of the bark) - broken into small, regular, rectangular pieces or flakes.
type - plant material on which the species is based.

umbel - the end cluster of buds, flowers or fruits, held by the peduncle, if present.

valve (of the fruit) - part of the fruit which is contracted or raised upon drying to allow release of the seeds.
var. - abbreviation of variety. Varieties of a species have less distinctiveness to one another than species do to one another, and usually do not occur in geographically different areas to the other variety(s) of that species.
venation - the pattern of the veins in the leaf.

Index of scientific names

Names printed in **boldface** indicate major entries

Index of common names

E. eucentrica on the Maralinga to Oak Valley road, Great Victoria Desert, S.A.